Android Design Patterns and Best Practice

Create reliable, robust, and efficient Android apps with
industry-standard design patterns

Kyle Mew

BIRMINGHAM - MUMBAI

Android Design Patterns and Best Practice

First published: December 2016

Production reference: 1221216

Published by Packt Publishing Ltd.
Livery Place
35 Livery Street
Birmingham
B3 2PB, UK.
ISBN 978-1-78646-721-8

www.packtpub.com

Credits

Author
Kyle Mew

Reviewer
Víctor Albertos

Commissioning Editor
Amarabha Banerjee

Acquisition Editor
Shweta Pant

Content Development Editor
Narendrakumar Tripathi

Technical Editor
Anushree Arun Tendulkar

Copy Editor
Safis Editing

Project Coordinator
Devanshi Doshi

Proofreader
Safis Editing

Indexer
Mariammal Chettiyar

Graphics
Jason Monteiro

Production Coordinator
Shantanu N. Zagade

About the Author

Kyle Mew has been programming since the early eighties and has written for several technology websites. He has also written three radio plays and three other books on Android development.

About the Reviewer

Victor Albertos has been developing Android apps for 5 years. During this time, he had the chance to explore several approaches to achieve maintainable and testable code. From an imperative style, gradually he has switch to a more declarative one, applying concepts and theories from functional languages such as Haskell to the Android and Java ecosystem.

He is an active open source contributor, sharing with the Android community useful libraries, mainly focused on reactive programming.

Currently he is working as an Android architect at Sabadell Bank.

> *I would like to thank to my family for their constant support, my few friends for being always there, and specially to Inma, who I deeply love.*

www.PacktPub.com

For support files and downloads related to your book, please visit www.PacktPub.com.

Did you know that Packt offers eBook versions of every book published, with PDF and ePub files available? You can upgrade to the eBook version at www.PacktPub.com and as a print book customer, you are entitled to a discount on the eBook copy. Get in touch with us at service@packtpub.com for more details.

At www.PacktPub.com, you can also read a collection of free technical articles, sign up for a range of free newsletters and receive exclusive discounts and offers on Packt books and eBooks.

https://www.packtpub.com/mapt

Get the most in-demand software skills with Mapt. Mapt gives you full access to all Packt books and video courses, as well as industry-leading tools to help you plan your personal development and advance your career.

Why subscribe?

- Fully searchable across every book published by Packt
- Copy and paste, print, and bookmark content
- On demand and accessible via a web browser

Customer Feedback

Thank you for purchasing this Packt book. We take our commitment to improving our content and products to meet your needs seriously—that's why your feedback is so valuable. Whatever your feelings about your purchase, please consider leaving a review on this book's Amazon page. Not only will this help us, more importantly it will also help others in the community to make an informed decision about the resources that they invest in to learn.

You can also review for us on a regular basis by joining our reviewers' club. **If you're interested in joining, or would like to learn more about the benefits we offer, please contact us**: customerreviews@packtpub.com

Table of Contents

Preface

Welcome to *Android Design Patterns and Best Practice*, a comprehensive guide to how to get the most out of your apps with the tried and tested programming philosophy, design patterns. These patterns provide a logical and elegant approach to solving many of the development problems that coders face. These patterns act as a guide creating a clear path from problem to solution, and although applying a design pattern does not guarantee best practice in itself, it will hugely assist the process and make the discovery of design flaws far easier. Design patterns can be implemented on very many platforms and written in as many programming languages. Some code libraries even apply patterns as part of their internal mechanics, and many readers will already be familiar with the Java Observer and Observable classes. The Android SDK we will be exploring makes great use of many patterns, such as factories, builders and listeners (which are really just observer patterns). Although we will cover these built-in design patterns, the book will mostly explore how we can build our own, custom made, patterns and apply them to Android development. Rather than approach each design pattern in turn, this book approaches the subject from the perspective of a developer, moving through each aspect of app development exploring individual design patterns as they would arise in the course of building an Android app. To clarify this journey, we will be focusing on a single imaginary app, designed to support a small business. This will take us from application conception through to publication, covering such topics as UI design, internal logic and user interaction along the way. During each of these steps we will explore those design patterns that are relevant to that process, by first exploring the pattern in its abstract form and then applying it to that particular situation. By the end of the book you will have learned how design patterns can be applied to all aspects of Android development and how using them assists best practice. It is the concept of design patterns that is more important than any specific pattern itself. Patterns can, and should, be adapted to suit our specific purposes, and by learning this way of approaching app development, we can even go on to create entirely original patterns of our own.

What this book covers

Chapter 1, *Design Patterns*, introduces the development environment, and two common design patterns, the factory and abstract factory.

Chapter 2, *Creational Patterns*, covers material and interface design, exploring the design support library and the builder design pattern.

Chapter 3, *Material Patterns*, introduces Android User Interfaces and some of the most significant material design components such as the app bar and sliding navigation drawer. This will introduce menus and action icons and how to implement them and how to use a drawer listener to detect user activity.

Chapter 4, *Layout Patterns*, leads in from the previous one, delving further into Android Layout design patterns and how gravity and weight can be used to create layouts that work on a variety of devices. This will take us onto how Android handles device orientation and screen size and shape differences. The strategy pattern is introduced and demonstrated.

Chapter 5, *Structural Patterns*, delves us deeper into the design library and create a layout governed by a coordinator layout with a recycler view. This requires exploring the adapter design pattern, first the internal versions and then we build one of our own, as well as a bridge pattern and facade and filter patterns.

Chapter 6, *Activating Patterns*, shows us how to apply patterns directly to our app. We cover more design library features such as collapsing toolbars, scrolling and divider. We crate a custom dialog, triggered by user activity. We revisit the factory patterns and show how a builder pattern can be used to inflate a UI.

Chapter 7, *Combining Patterns*, introduces and demonstrates two new structural patterns the prototype and decorator, covering their flexibility. This is then put in to practice as we use the patterns to control a UI comprised of different compound buttons such as switches and radio groups.

Chapter 8, *Composing Patterns*, concentrates on the composite pattern and how it can be used in many situations and how to select the right situation. We then continue to use it in a practical demonstration to inflate a nested UI. This leads into the storage and retrieval of persistent data, using internal storage, application files and ultimately user settings in the form of shared preferences.

Chapter 9, *Observing Patterns*, looks at the visual processes involved in the transitions from one Activity to another and how this can be far more than mere decoration. The reader learns how to apply transitions and shared elements to efficiently use the minimal screen space of mobile devices and simplify application use and operation.

Chapter 10, *Behavioral Patterns*, concentrates on the major behavioral patterns, template, strategy, visitor and state. It provides working demonstrations of each and covers their flexibility and usage.

Chapter 11, *Wearable Patterns*, shows working of Android Wear, TV and Auto, demonstrating how to set up and configure each in turn. We examine the differences between these and the standard handheld app.

Chapter 12, *Social Patterns*, shows how to add web functionality and social media. Firstly the WebView is explored and how it can be used to create internal web-apps. Next, how to connect our app to Facebook is explored, showing how this is done and what we can do with it. The chapter concludes by examining other social platforms, such as Twitter.

Chapter 13, *Distribution Patterns*, covers deployment, publication and monetization of Android apps. The reader is led through the registration and publication process and we take a look at advertising options and which are best suited to which purpose. Finally we look at how we can maximize potential users with a few tips and tricks of deployment.

What you need for this book

Android Studio and SDK are both open source and can be installed from a single package. With one minor exception, which is outlined thoroughly in the relevant chapter, this is all the software required for this book.

Who this book is for

This book is intended for Android developers who have some basic android development experience. Basic Java programming knowledge is a must to get the most out of this book.

Conventions

In this book, you will find a number of text styles that distinguish between different kinds of information. Here are some examples of these styles and an explanation of their meaning.

Code words in text, database table names, folder names, filenames, file extensions, pathnames, dummy URLs, user input, and Twitter handles are shown as follows: "Add three TextViews to your layout, and then add the code to your MainActivity's onCreate() method."

A block of code is set as follows:

```
Sequence prime = (Sequence) SequenceCache.getSequence("1");
primeText.setText(new StringBuilder()
        .append(getString(R.string.prime_text))
        .append(prime.getResult())
        .toString());
```

When we wish to draw your attention to a particular part of a code block, the relevant lines or items are set in bold:

```
@Override
public String getDescription() {
    return filling.getDescription() + " Double portion";
}
```

Any command-line input or output is written as follows:

```
/gradlew clean:
```

New terms and **important words** are shown in bold. Words that you see on the screen, for example, in menus or dialog boxes, appear in the text like this: "Enable developer options on your handset. On some models, this can involve navigating to **Settings** | **About phone**"

Warnings or important notes appear in a box like this.

Tips and tricks appear like this.

Reader feedback

Feedback from our readers is always welcome. Let us know what you think about this book-what you liked or disliked. Reader feedback is important for us as it helps us develop titles that you will really get the most out of. To send us general feedback, simply e-mail feedback@packtpub.com, and mention the book's title in the subject of your message. If there is a topic that you have expertise in and you are interested in either writing or contributing to a book, see our author guide at www.packtpub.com/authors.

Customer support

Now that you are the proud owner of a Packt book, we have a number of things to help you to get the most from your purchase.

Downloading the example code

You can download the example code files for this book from your account at `http://www.p acktpub.com`. If you purchased this book elsewhere, you can visit `http://www.packtpub.c om/support` and register to have the files e-mailed directly to you.

You can download the code files by following these steps:

1. Log in or register to our website using your e-mail address and password.
2. Hover the mouse pointer on the **SUPPORT** tab at the top.
3. Click on **Code Downloads & Errata**.
4. Enter the name of the book in the **Search** box.
5. Select the book for which you're looking to download the code files.
6. Choose from the drop-down menu where you purchased this book from.
7. Click on **Code Download**.

Once the file is downloaded, please make sure that you unzip or extract the folder using the latest version of:

- WinRAR / 7-Zip for Windows
- Zipeg / iZip / UnRarX for Mac
- 7-Zip / PeaZip for Linux

The code bundle for the book is also hosted on GitHub at `https://github.com/PacktPubl ishing/Android-Design-Patterns-and-Best-Practice`. We also have other code bundles from our rich catalog of books and videos available at `https://github.com/PacktPublish ing/`. Check them out!

Downloading the color images of this book

We also provide you with a PDF file that has color images of the screenshots/diagrams used in this book. The color images will help you better understand the changes in the output. You can download this file from `https://www.packtpub.com/sites/default/files/down loads/AndroidDesignPatternsandBestPractice.pdf`.

Errata

Although we have taken every care to ensure the accuracy of our content, mistakes do happen. If you find a mistake in one of our books-maybe a mistake in the text or the code-we would be grateful if you could report this to us. By doing so, you can save other readers from frustration and help us improve subsequent versions of this book. If you find any errata, please report them by visiting `http://www.packtpub.com/submit-errata`, selecting your book, clicking on the **Errata Submission Form** link, and entering the details of your errata. Once your errata are verified, your submission will be accepted and the errata will be uploaded to our website or added to any list of existing errata under the Errata section of that title.

To view the previously submitted errata, go to `https://www.packtpub.com/books/content/support` and enter the name of the book in the search field. The required information will appear under the **Errata** section.

Piracy

Piracy of copyrighted material on the Internet is an ongoing problem across all media. At Packt, we take the protection of our copyright and licenses very seriously. If you come across any illegal copies of our works in any form on the Internet, please provide us with the location address or website name immediately so that we can pursue a remedy.

Please contact us at `copyright@packtpub.com` with a link to the suspected pirated material.

We appreciate your help in protecting our authors and our ability to bring you valuable content.

Questions

If you have a problem with any aspect of this book, you can contact us at `questions@packtpub.com`, and we will do our best to address the problem.

1
Design Patterns

Design patterns have long been considered some of the most reliable and useful approaches to solving common software design problems. Patterns provide general and reusable solutions to frequently occurring development issues, such as how to add functionality to an object without changing its structure or how best to construct complex objects.

There are several advantages to applying patterns, not least the way that this approach assists the developer in following best practices and how it simplifies the management of large projects. These benefits are achieved by providing overall software structures (patterns) that can be reused to solve similar problems. This is not to say that code can be simply cut and pasted from one project to another but that the concepts themselves can be used over and over again in many different situations.

There are a good many other benefits to applying programming patterns, all of which will be covered at some point in the book, but here are one or two that are worthy of mention now:

- Patterns provide an efficient common language between developers working in teams. When a developer describes a structure as say, an **adapter** or a **facade**, other developers will understand what this means and will immediately recognize the structure and purpose of the code.
- The added layers of abstraction that patterns provide make modifications and alterations to code that is already under development much easier. There are even patterns designed for these very situations.
- Patterns can be applied at many scales, from the overall architectural structure of a project right down to the manufacturing of the most basic object.
- The application of patterns can vastly reduce the amount of inline commentary and general documentation required, as patterns also act as their own description. The name of a class or interface alone can explain its purpose and place within a pattern.

The Android development platform lends itself nicely to the employment of patterns, as not only are applications created largely in Java, but the SDK contains many APIs that make use of patterns themselves, such as **factory** interfaces for creating objects and **builders** for constructing them. Simple patterns such as **Singletons** are even available as a template class type. In this book, we shall see not only how to put together our own, large-scale patterns but also how make use of these built-in structures to encourage best practice and simplify coding.

In this chapter, we will begin by taking a brief look at how the book as a whole will pan out, the patterns we will be using, the order in which we will approach them, and the demonstration app that we will be building to see how patterns can be applied in real-world situations. This will be followed by a quick examination of the SDK and which components will best assist us on our journey, in particular, the role that the **support library** provides, enabling us to develop for many platform versions at once. There is no better way to learn than actual experience, and so the rest of the chapter will be taken up by developing a very simple demonstration app and employing our first pattern, the **factory pattern**, and its associated **abstract factory** pattern.

In this chapter, you will learn the following:

- How patterns are categorized and which patterns are covered here
- The purpose of the book's demonstration app
- How to target platform versions
- What support libraries do
- What a factory pattern is and how to construct one
- How to follow a UML class diagram
- How to test an app on both real and virtual devices
- How to monitor an app during runtime
- How to use simple debugging tools to test code
- What an abstract factory pattern is and how to use one

How this book works

The purpose of this book is to show how the application of design patterns can directly assist the development of Android applications. During the course of the book we will be concentrating on the development of a complete client-side mobile application, focusing particularly on when, why, and how patterns can and should be employed during Android development.

Historically, there has been a certain amount of disagreement as to what exactly constitutes a pattern. However, the 23 patterns laid out in the 1994 book *Design Patterns* by Erich Gamma, Richard Helm, Ralph Johnson, and John Vlissides, the so-called gang of four, are widely accepted as the definitive set and provide solutions to nearly all of the software engineering problems that we are likely to encounter, and it is for this reason that these patterns will form the backbone of the book. These patterns can be broken down into three categories:

- **Creational** – Used for creating objects
- **Structural** – Used for organizing groups of objects
- **Behavioral** – Used for communication between objects

The practical nature of this book means that we will not be tackling these categories in the order they appear here; instead, we will explore individual patterns as and when they arise naturally during the development of our app, although this generally means starting with creating a structure first.

It would be difficult, clumsy, and unrealistic to incorporate all design patterns into a single application, so here will attempt to apply as many as might seem realistic. For those patterns we decide not to use directly, we will at least explore how we might have done so, and in every case give at least one practical example of how they are used.

Patterns are not written in stone, nor do they solve every possible problem, and towards the end of the book we will look at how, once we have a grasp of the subject, we can create our own patterns or adapt extant ones to suit the rare circumstances where none of the established patterns fit.

In short, patterns are not a set of rules but rather a series of well-worn paths leading from known problems to tested solutions. If you spot a shortcut along the way, then by all means use it. If you stick to it consistently, then you will have created a pattern of your own that is just as valid as the traditional ones that we will cover here.

The first few chapters of the book concentrate on UI design and introduce a few essential design patterns and how they work conceptually. From about Chapter 6, *Activating Patterns*, onward, we will start to apply these and other patterns to real-world examples, and to one application in particular. The final chapters concentrate on the final stages of development, such as adapting applications for differing devices, a task almost purpose-built for design patterns, reaching the widest possible market and how to monetize our apps.

In case you are new to Android development, the instructions laid out in the first two or three chapters are given in detail. If you are already familiar with Android development, you will be able to skip these sections and concentrate on the patterns themselves.

Before we get stuck into our first patterns, it makes sense to take a closer look at the app we will be building during the course of the book, as well as the challenges and opportunities it presents.

What we will build

As mentioned earlier, throughout the course of this book, we will be building a small but complete Android application. It will be a good idea now to take a little look at what we will be building and why.

We will put ourselves in the position of an independent Android developer that has been approached by a potential client that runs a small business making and delivering fresh sandwiches to several local office buildings. There are several issues facing our client that they believe can be solved with a mobile app. To see what solutions an app might provide, we will break the situation down into three sections: the scenario, the problem, and the solution.

The scenario

The client runs a small but successful business making and then delivering fresh sandwiches to nearby office workers so that they can buy and eat them at their desks. The sandwiches are very good and, as a result of word-by-mouth advertising, are growing in popularity. There is a good opportunity for the business to expand, but there are some glaring inefficiencies in the business model that the client believes can be resolved with the use of a mobile app.

The problem

It is almost impossible for the client to anticipate demand. There are many occasions where too many of a particular sandwich are made, leading to wastage. Likewise, there are times where insufficient sandwich lines are prepared, leading to a loss in sales. Not only this, but the word-of-mouth advertising the customers provide limits the expansion of the business to a small geographical area. The client has no reliable way of knowing if it is worth investing in more staff, a motorbike to travel further afield, or even whether to open new kitchens in other parts of town.

The solution

A mobile app, provided free for all customers, not only solves these problems but makes available a whole new set of opportunities. Never mind that an app will solve the issues of unanticipated demand; we now have the chance to take this to a whole new level. Why just present the customer with a set menu when we can offer them the chance to construct their own personalized sandwich from a list of ingredients? Maybe they love the cheese and pickle sandwich our client already makes but fancy it with a slice or two of apple, or prefer mango chutney to pickle. Maybe they are vegetarian and prefer to filter out meat products from their choices. Maybe they have allergies. All of these needs can be met with a well-designed mobile app.

Furthermore, the geographical limitations of word-of-mouth advertising, and even local promotions such a as billboards or notices in local papers, gives no indication of just how successful a business might be on a larger stage. The use of social media, on the other hand, can give our client clear insights into current trends as well as spread the word to the widest possible audience.

Not only can our client now judge accurately the scope of their business but can also add entirely new features unique to the digital nature of modern life, such as the gamification of the app. Competitions, puzzles, and challenges can provide a whole new dimension to engaging customers and present a powerful technique to increasing revenue and market presence.

With the task ahead now a little clearer, we are now in a position to start coding. We will start with a very simple demonstration of the factory pattern, and on the way take a closer look at some of the features of the SDK that we will be finding useful along the way.

Targeting platform versions

To keep up with the latest technology, new versions of the Android platform are released frequently. As developers, this means we can incorporate the newest features and developments into our applications. The obvious drawback to this is the fact that only the very newest devices will be able to run this platform and these devices represent only a tiny proportion of the entire market. Take a look at this chart taken from the developer dashboard:

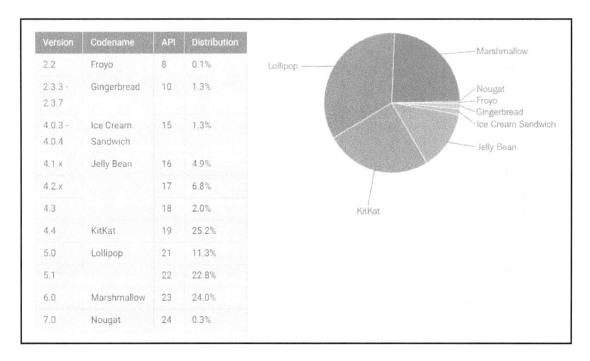

Version	Codename	API	Distribution
2.2	Froyo	8	0.1%
2.3.3 - 2.3.7	Gingerbread	10	1.3%
4.0.3 - 4.0.4	Ice Cream Sandwich	15	1.3%
4.1.x	Jelly Bean	16	4.9%
4.2.x		17	6.8%
4.3		18	2.0%
4.4	KitKat	19	25.2%
5.0	Lollipop	21	11.3%
5.1		22	22.8%
6.0	Marshmallow	23	24.0%
7.0	Nougat	24	0.3%

The dashboard can be found at `developer.android.com/about/dashboards/index.html` and contains this and other up-to-date information that is very useful when first planning a project.

As you can see, the vast majority of Android devices still run on older platforms. Fortunately, Android makes it possible for us to target these older devices while still being able to incorporate features from the most recent platform version. This is largely achieved through the use of the **support library** and by setting a minimum SDK level.

Deciding which platforms to target is one of the first decisions we will need to take, and although it is possible to change this at a later date, deciding early which features to incorporate and knowing how these will appear on older devices can greatly simplify the overall task.

To see how this is done, start a new Android Studio project, call it anything you choose, and select **Phone and Tablet** as the form factor and **API 16** as the **Minimum SDK**.

From the list of templates, select **Empty Activity** and leave everything else as is.

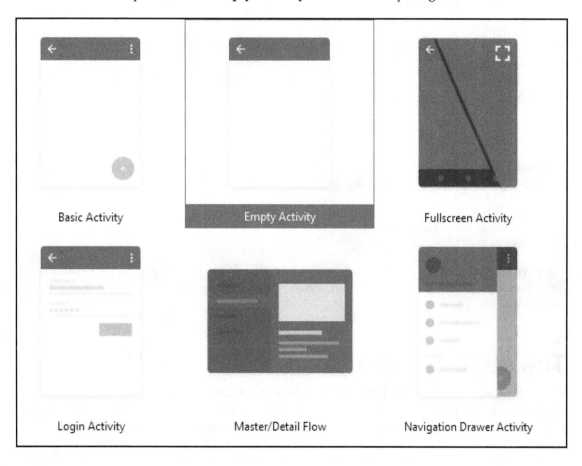

| Basic Activity | Empty Activity | Fullscreen Activity |
| Login Activity | Master/Detail Flow | Navigation Drawer Activity |

Android Studio will automatically select the highest available SDK version as the target level. To see how this is applied, open the `build.gradle` (Module: app) file from the project pane and note the `defaultConfig` section, which will resemble the following code:

```
defaultConfig {
    applicationId "com.example.kyle.factoryexample"
    minSdkVersion 16
    targetSdkVersion 25
    versionCode 1
    versionName "1.0"
}
```

This ensures that our project will compile correctly for this range of API levels, but if we were building an app that we intended to publish, then we would need to inform the Google Play store which devices to make our app available on. This can be done with the `build.gradle` modular file, like so:

```
minSdkVersion 21
targetSdkVersion 24
```

We would also need to edit `AndroidManifest.xml` file. For the example here, we would add the following `uses-sdk` element to the `manifest` node:

```
<uses-sdk
    android:minSdkVersion="16"
    android:targetSdkVersion="25" />
```

Once we have determined the range of platforms we wish to target, we can get on and see how the support library allows us to incorporate many of the latest features on many of the oldest devices.

The support library

When it comes to building backwards-compatible applications, the support library is undoubtedly our most powerful tool. It is in fact a series of individual code libraries that work by providing alternative classes and interfaces to those found in the standard APIs.

There are around 12 individual libraries and they do not only provide compatibility; they also include common UI components such as sliding drawers and floating action buttons that would otherwise have to be built from scratch. They can also simplify the process of developing for different screen sizes and shapes, as well as adding one or two miscellaneous functions.

As we are developing with Android Studio, we should download the **support repository** rather than the support library as the repository is designed specifically for the studio, provides exactly the same functionality, and is more efficient.

In the example we are working on in this chapter, we will not be making any use of support libraries. The only one the project includes is the v7 appcompat library, which was added automatically for us when we started the project. We will be returning to support libraries often in the book, so for now, we can concentrate on applying our first pattern.

The factory pattern

The factory pattern is one of the most widely used creational patterns. As its name suggests, it makes things, or more precisely, it creates objects. Its usefulness lies in the way it uses a common interface to separate logic from use. The best way to see how this works is simply to build one now. Open the project we began a page or two previously, or start a new one. Minimum and target SDK levels are not important for this exercise.

 Selecting an API level of 21 or higher allows Android Studio to employ a technology known as hot-swapping. This avoids having to completely rebuild a project each time it is run and vastly speeds up the testing of an app. Even if you intend to finally target a lower platform, the time hot-swapping saves makes it well worth your while lowering this target once the app is as good as developed.

We are going to build a very simple example app that generates objects to represent the different types of bread our sandwich builder app might offer. To emphasize the pattern, we will keep it simple and have our objects return nothing more sophisticated than a string:

1. Locate the MainActivity.java file in the project view.
2. Right-click it and create a New | Java Class of **KindInterface** called Bread:

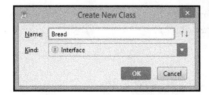

3. Complete the interface as follows:

```
public interface Bread {

    String name();
    String calories();
}
```

4. Create concrete classes of Bread, like so:

```
public class Baguette implements Bread {

    @Override
    public String name() {
        return "Baguette";
    }

    @Override
    public String calories() {
        return " : 65 kcal";
    }
}

    public class Roll implements Bread {

    @Override
    public String name() {
        return "Roll";
    }
    @Override
    public String calories() {
        return " : 75 kcal";
    }
}

    public class Brioche implements Bread {

    @Override
    public String name() {
        return "Brioche";
    }

    @Override
    public String calories() {
        return " : 85 kcal";
    }
}
```

5. Next, create a new class called `BreadFactory` that looks like this:

```
public class BreadFactory {

    public Bread getBread(String breadType) {

        if (breadType == "BRI") {
            return new Brioche();

        } else if (breadType == "BAG") {
            return new Baguette();

        } else if (breadType == "ROL") {
            return new Roll();
        }

        return null;
    }
}
```

UML diagrams

The key to understanding design patterns lies in understanding their structure and how component parts relate to each other. One of the best ways to view a pattern is pictorially, and the Unified Modeling Language (UML) class diagrams are a great way to accomplish this.

Consider the pattern we just created expressed diagrammatically, like so:

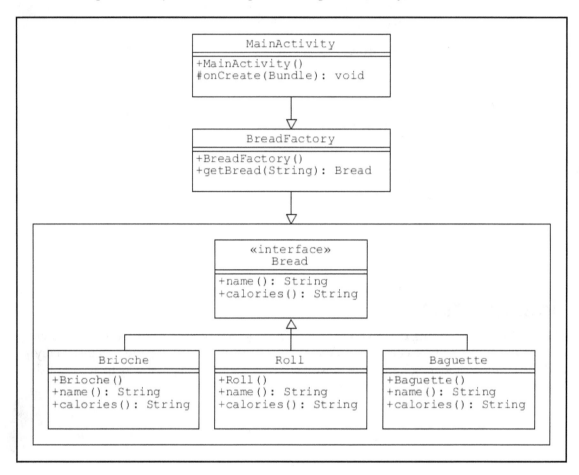

With our pattern in place, all that is required is to see it in action. For this demonstration, we will make use of the **TextView** in our layout that the template generated for us and the onCreate() method that is called every time our main activity is started:

1. Open the activity_main.xml file in **Text** mode.
2. Add an id to the text view, like so:

```
<TextView
    android:id="@+id/text_view"
    android:layout_width="match_parent"
    android:layout_height="wrap_content" />
```

3. Open the `MainActivity.java` file and edit the `onCreate()` method to match the following code:

```
@Override
protected void onCreate(Bundle savedInstanceState) {
    super.onCreate(savedInstanceState);
    setContentView(R.layout.activity_main);

    TextView textView = (TextView) findViewById(R.id.text_view);

    BreadFactory breadFactory = new BreadFactory();
    Bread bread = breadFactory.getBread("BAG");

    textView.setText(new StringBuilder()
            .append(bread.name())
            .toString());
}
```

Depending on how you have Android Studio set up, you may have to import the TextView widget: `import android.widget.TextView;`. Usually, the editor will prompt you and import the widget with a simple press of Alt + Enter.

You can now test the pattern on an emulator or real device:

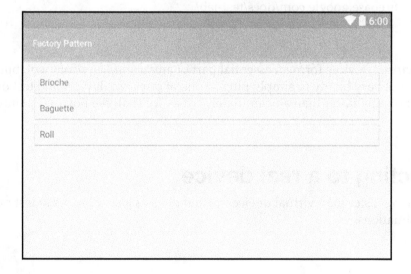

This may appear at first glance as an incredibly long-winded way to achieve a very simple goal, but therein lies the beauty of patterns. The added layers of abstraction allow us to modify our classes without having to edit our activity and vice versa. This usefulness will become more apparent as we develop more complex objects and encounter situations that require more than a single factory.

The example we created here is too simple to really require any testing, but now is as good a time as any to explore how we test Android apps on both real and virtual devices, as well as how we can monitor performance and use debugging tools to test output without having to add unnecessary screen components.

Running and testing an app

There are an enormous number of Android devices available today, and they come in a huge variety of shapes and sizes. As developers, we want our applications to run on as many devices and form factors as possible, and we want to be able to do it with the minimum of coding. Fortunately, the Android platform is well suited to this challenge, allowing us to easily adapt layouts and to construct virtual devices to match any form factor we can imagine.

 Google provide a very handy cloud-based app testing facility at firebase.google.com/docs/test-lab/

Obviously, virtual devices form an essential part of any testing environment, but that is not to say that is not very handy to simply plug in one of our own devices and test our apps on that. This is not only faster than any emulator, but as we shall see now, very simple to set up.

Connecting to a real device

As well as being faster than virtual devices, actual devices also allow us to test our apps in real-world situations.

Connecting a real device to our development environment requires two steps:

1. Enable developer options on your handset. On some models, this can involve navigating to `Settings | About phone` and tapping on `Build number` seven times, which will then add `Developer options` to your settings. Use this to enable **USB debugging** and to **Allow mock locations**.

2. It is more than likely that you will now be able to connect your device to your workstation via a USB or the WiFi plugin cable and for it to show up when you open Android Studio. If not, you may need to open the SDK Manager and install the **Google USB driver** from the **Tools** tab. On some rare occasions, you may have to download a USB driver from the device's manufacturer.

A real device can be very useful for quickly testing changes in the functionality of an app, but to develop how our app looks and behaves on a variety of screen shapes and sizes will mean that we will create some virtual devices.

Connecting to a virtual device

Android virtual devices (AVDs) allow developers to experiment freely with a variety of emulated hardware setups, but they are notoriously slow, can exhaust the resources of many computer systems, and lack many of the features present in actual devices. Despite these drawbacks, virtual devices are an essential part of an Android developers' toolkit, and by taking a few things into consideration, many of these hindrances can be minimized:

- Strip your virtual device down to only the features your app is designed for. For example, if you are not building an app that takes photos, remove camera functionality from the emulator; one can always be added later.
- Keep the memory and storage requirements of the AVD to a minimum. Another device can easily be created as and when an app needs it.
- Only create AVDs with very recent API levels when you need to test for specifically new features.
- Begin by testing on virtual devices with low screen resolution and density. These will run much faster and still allow you to test for different screen sizes and aspect ratios.
- Try separating very resource demanding functions and testing them individually. For example, if your app utilizes large collections of high-definition images, you can save time by testing this functionality separately.

It is generally quicker to construct virtual devices to suit specific purposes rather than an all-purpose one to test all our apps on, and there is a growing number of third-party Android emulators available, such as *Android-x86* and *Genymotion*, that are often faster and have more development features.

It is also worth noting that when testing for layouts only, Android Studio provides some powerful preview options that allow us to view our potential UIs on a wide number of form factors, SDK levels and themes, as can be seen in the next image:

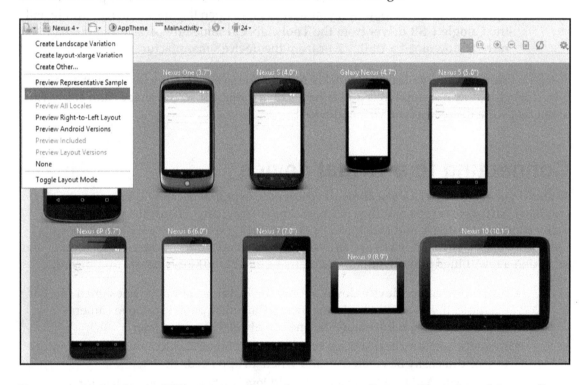

For now, create a basic AVD to run and test the current project on. There is nothing really to test, but we are going to see how to monitor our app's behavior during runtime and how to use the debug monitor service to test output without having to use the device screen, which is not an attractive way to debug a project.

Monitoring devices

The following demonstration works equally well on either an emulated device or a real one, so select whichever is easiest for you. If you are creating an AVD, then it will not require a large or high density screen or a large memory:

1. Open the project we just worked on.
2. From the `Tools | Android` menu, enable **ADB Integration**.

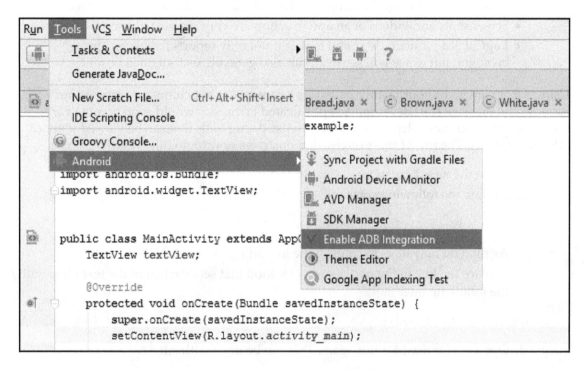

3. From the same menu, select **Android Device Monitor**, although this may well already be running.
4. Now run the application on your connected device with the Android Monitor.

The Device Monitor is useful in several ways:

- The **Monitors** tab can be used during runtime to view live system information such as how much memory or CPU time our app is using. This can be particularly helpful when we want to see what resources our apps are using when they are not running in the foreground.
- The monitor can be set to collect a variety of data, such as method tracking and resource usage, and store these as files, which can be viewed in the **Captures** pane (which can usually be opened from the left-hand gutter).
- Screenshots and videos of an app in action are very simple to capture.
- **LogCat** is a particularly useful tool as it not only reports live on an app's behavior, but as we will see next, can also generate user-defined output.

Using a text view to test our factory pattern is a convenient but clumsy way to test our code for now, but once we start developing sophisticated layouts, it would soon become very inconvenient. A far more elegant solution is to use debug tools that can be viewed without affecting our UIs. The rest of this exercise demonstrates how to do this:

1. Open the `MainActivity.java` file.
2. Declare the following constant:

```
private static final String DEBUG_TAG = "tag";
```

3. Again, you may have to confirm the importing of `android.util.Log;`.
4. Replace the line in the `onCreate()` method that sets the text of the text view with the following line:

```
Log.d(DEBUG_TAG, bread);
```

5. Open the Device Monitor again. This can be done with Alt + 6.

6. From the dropdown in the top-right of the monitor, select **Edit Filter Configuration**.

7. Complete the resultant dialog, as seen here:

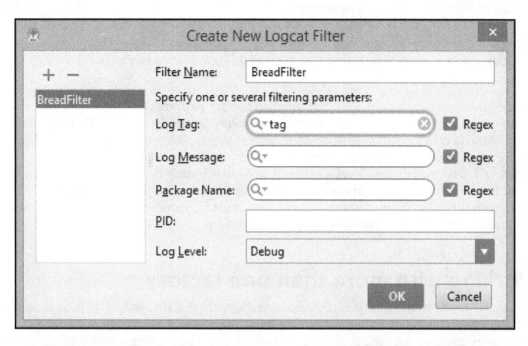

Running the app and testing our factory demo should produce an output in the logcat monitor similar to the one seen here:

```
05-24 13:25:52.484 17896-17896/? D/tag: Brioche
05-24 13:36:31.214 17896-17896/? D/tag: Baguette
05-24 13:42:45.180 17896-17896/? D/tag: Roll
```

You can, of course, still use System.out.println() if you like, and it will print out in the ADB monitor, but you will have to search for it among the other output.

Having seen how we can test our apps on both real and virtual devices and how we can use debug and monitoring tools to interrogate our apps during runtime, we can now move on to a more realistic situation involving more than a single factory and an output more sophisticated than a two-word string.

The abstract factory pattern

When making a sandwich, bread is only our first and most basic ingredient; we obviously need some kind of filling. In programming terms, this could mean simply building another interface like `Bread` but calling it `Filling` and providing it with its own associated factory. Equally, we could create a global interface called `Ingredient` and have both `Bread` and `Filling` as examples of this. Either way, we would have to do a fair bit of re-coding elsewhere.

The design pattern paradigm offers the **abstract factory pattern** as perhaps the most adaptable solution to this dilemma. An abstract factory is simply **a factory that creates other factories**. The added layer of abstraction that this requires is amply paid off when we consider how little the top-level control code in our main activity needs to be altered, if at all. Being able to modify low-level structures without affecting those preceding constitutes one of the major reasons for applying design patterns, and when applied to complex architectures, this flexibility can shave many weeks off development time and allow more room for experimentation than other approaches.

Working with more than one factory

The similarities between this next project and the last are striking, as they should be; one of the best things about patterns is that we can reuse structures. You can either edit the previous example or start one from scratch. Here, we will be starting a new project; hopefully that will help make the pattern itself clearer.

The **abstract factory** works in a slightly different way to our previous example. Here, our activity makes uses of a factory generator, which in turn makes use of an abstract factory class that handles the task of deciding which actual factory to call, and hence which concrete class to create.

As before we will not concern ourselves with the actual mechanics of input and output, but rather concentrate on the pattern's structure. Before continuing, start a new Android Studio project. Call it whatever you choose, set the minimum API level as low as you like, and use the Blank Activity template:

1. We begin, just as we did before, by creating the interface; only this time, we will need two of them: one for the bread and one for the filling. They should look like this:

```
public interface Bread {

    String name();
```

```
      String calories();
}

public interface Filling {

      String name();
      String calories();
}
```

2. As before, create concrete examples of these interfaces. Here, to save space, we will just create two of each. They are all almost identical, so here is just one:

```
public class Baguette implements Bread {

      @Override
      public String name() {
          return "Baguette";
      }

      @Override
      public String calories() {
          return " : 65 kcal";
      }
}
```

3. Create another `Bread` called `Brioche` and two fillings called `Cheese` and `Tomato`.

4. Next, create a class that can call on each type of factory:

```
public abstract class AbstractFactory {

      abstract Bread getBread(String bread);
      abstract Filling getFilling(String filling);
}
```

5. Now create the factories themselves. First, `BreadFactory`:

```
public class BreadFactory extends AbstractFactory {

      @Override
      Bread getBread(String bread) {

          if (bread == null) {
              return null;
          }

          if (bread == "BAG") {
```

```
                return new Baguette();
            } else if (bread == "BRI") {
                return new Brioche();
            }

            return null;
        }

        @Override
        Filling getFilling(String filling) {
            return null;
        }
    }
```

6. And then, `FillingFactory`:

```
public class FillingFactory extends AbstractFactory {

        @Override
        Filling getFilling(String filling) {

            if (filling == null) {
                return null;
            }

            if (filling == "CHE") {
                return new Cheese();
            } else if (filling == "TOM") {
                return new Tomato();
            }

            return null;
        }

        @Override
        Bread getBread(String bread) {
            return null;
        }
    }
```

7. Finally, add the factory generator class itself:

```
public class FactoryGenerator {

        public static AbstractFactory getFactory(String factory) {

            if (factory == null) {
                return null;
```

```
        }

        if (factory == "BRE") {
            return new BreadFactory();
        } else if (factory == "FIL") {
            return new FillingFactory();
        }

        return null;
    }
}
```

8. We can test our code just as before, with a debug tag, like so:

```
AbstractFactory fillingFactory = FactoryGenerator.getFactory("FIL");
Filling filling = fillingFactory.getFilling("CHE");
Log.d(DEBUG_TAG, filling.name()+" : "+filling.calories());

AbstractFactory breadFactory = FactoryGenerator.getFactory("BRE");
Bread bread = breadFactory.getBread("BRI");
Log.d(DEBUG_TAG, bread.name()+" : "+bread.calories());
```

When tested, this should produce the following output in the Android monitor:

```
com.example.kyle.abstractfactory D/tag: Cheese :   : 155 kcal
com.example.kyle.abstractfactory D/tag: Brioche :  : 85 kcal
```

By the time we reach the end of the book, each ingredient will be a complex object in its own right, with associated imagery and descriptive text, price, calorific value, and more. This is when adhering to patterns will really pay off, but a very simple example like the one here is a great way to demonstrate how creational patterns such as the abstract factory allow us to make changes to our products without affecting client code or deployment.

As before, our understanding of the pattern can be enhanced with a visual representation:

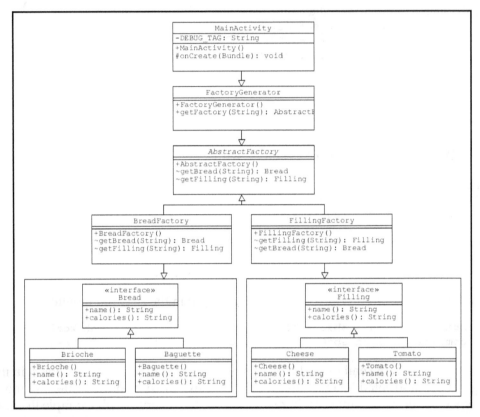

Imagine we wanted to include soft drinks in our menu. These are neither bread nor filling, and we would need to introduce a whole new type of object. The pattern of how to add this is already laid out. We will need a new interface that would be identical to the others, only called `Drink`; it would utilize the same `name()` and `calories()` methods, and concrete classes such as `IcedTea` could be implemented along exactly the same lines as above, for example:

```java
public class IcedTeaimplements Drink {

    @Override
    public String name() {
        return "Iced tea";
    }

    @Override
    public String calories() {
```

```
        return " : 110 kcal";
    }
}
```

We would need to extend our abstract factory with something like this:

```
abstract Drink getDrink(String drinkType);
```

We also, of course, need to implement a `DrinkFactory` class, but this too would have the same structure as the other factories.

In other words, we can add, delete, change, and generally muck around with the nuts and bolts of a project, without ever really having to bother with how these changes are perceived by the higher-level logic of our software.

The factory pattern is one of the most frequently used of all patterns. It can and should be used in many situations. However, like all patterns, it can be overused or underused, if not thought about carefully. When considering the overall architecture of a project, there are, as we shall see, many other patterns at our disposal.

Summary

We have covered quite a lot, considering that this is an introductory chapter. We've built examples of two of the best known and most useful design patterns and hopefully seen why they might be of use to us.

We began by looking at what patterns are and why we might use them in an Android setting. This was helped by taking a look at the development tools at our disposal, and how and why we can and should target specific platform versions and form factors.

We then applied this knowledge to create two very simple apps that employed basic factory patterns and saw how we can test and then retrieve data from an app running on any device, be it real or virtual.

This puts us in a great situation to take a look at other patterns and consider which to use when building a fully working app. This is something we will look at more closely in the next chapter, where we will introduce the builder pattern and how Android layouts are produced.

2
Creational Patterns

In the previous chapter, we took a look at the **factory pattern** and its associated **abstract factory pattern**. However, we looked at these patterns in quite a general way and not at how these objects, once created, can be represented and manipulated on an Android device. In other words, the patterns we built could have been applied in many other software environments, and to see how to make them more Android-specific we need to take a look at Android UIs and how they are composed.

In this chapter, we will concentrate on how to represent our products as Android UI components. We will use the **card view** to display these, and each card will contain a title, an image, some descriptive text, and the ingredient's calorific value, as can be seen in the following screenshot:

This will lead us to take an initial look at **material design,** a powerful and increasingly popular **visual design language** for creating clean and intuitive UIs. Conceived originally for the smaller screens of mobile devices, material design is now widely considered such a valuable UI paradigm that its use has spread from Android devices to websites and even other mobile platforms.

Material design is more than just fashionable, it provides a very effective series of guidelines for following best UI construction practices. Material design provides visual patterns that are analogous to the programmatic patterns we have already discussed. These patterns provide well-defined structures that are clean and simple to operate. Material design covers concepts such as proportion, scaling, typography, and spacing, all of which are very easily managed within the IDE and neatly prescribed by material design guidelines.

Once we have seen how to represent our ingredients as workable UI components, we will take a look at another commonly used creational pattern, the **builder pattern**. This will demonstrate a pattern that allows us to build a single *sandwich* object from our individual *ingredient* objects.

In this chapter, you will learn how to do the following:

- Edit material styles and themes
- Apply palettes
- Customize text settings
- Manage screen densities
- Include the card view support library
- Understand z-depth and shading
- Applying material design to a card view
- Create a builder pattern

Although it can be changed at any time, one of the first things we should consider when building an Android app is the color scheme. This is the way that the framework allows us to customize the color and appearance of many familiar screen components, such as the title and status bar background colors and text and highlight shades.

Applying themes

As developers, we want our apps to stand out from the crowd, but we also want to incorporate all the features that Android users are familiar with. One way to do this is by applying a particular color scheme throughout an app. This is most easily done by customizing or creating Android themes

Since API level 21 (Android 5.0), the **material theme** has been default on Android devices. It is, however, more than just a new look. The material theme also provides as default the touch feedback and transition animations that we associate with material design. As with all Android themes, the material themes are based on Android styles.

An **Android style** is a set of graphical properties defining the appearance of a particular screen component. Styles allow us to define everything from font size and background color to padding and elevation, and much more. An Android theme is simply a style applied across a whole activity or application. Styles are defined as XML files, and stored in the resources (`res`) directory of Android Studio projects.

Fortunately, Android Studio comes with a graphical **theme editor** that generates the XML for us. Nevertheless, it is always good to understand what is going on under the hood, and this is best seen by opening the abstract factory project from the last chapter or by starting a new one. From the project explorer, open the `res/values/styles.xml` file. It will contain the following style definition:

```xml
<style name="AppTheme" parent="Theme.AppCompat.Light.DarkActionBar">

    <item name="colorPrimary">@color/colorPrimary</item>
    <item name="colorPrimaryDark">@color/colorPrimaryDark</item>
    <item name="colorAccent">@color/colorAccent</item>

</style>
```

Here, only three colors are defined, although we could have had more, such as primary and secondary text colors, window background color, and others. The colors themselves are defined in the `colors.xml` file, which is also found in the `values` directory and will contain the following definitions:

```xml
<color name="colorPrimary">#3F51B5</color>
<color name="colorPrimaryDark">#303F9F</color>
<color name="colorAccent">#FF4081</color>
```

It is quite possible to apply more than one theme and incorporate as many styles as we might like, but generally speaking, a single theme applied across an entire application that customizes one of the default material themes is the easiest and cleanest answer.

The simplest way to customize the default theme is with the theme editor, which can be opened from the `Tools | Android` menu. The editor provides a powerful WYSIWYG preview pane that allows us to instantly view any changes we make as we make them, like so:

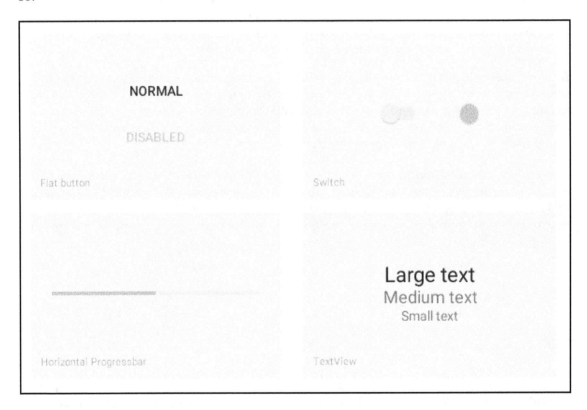

Although we are free to choose any colors we like for our theme, the material design guidelines are quite clear about how colors should be used together. This is best explained by taking a look at **material palettes**.

Customizing color and text

The first things we need to consider when applying a theme are colors and text. Material design guidelines recommend selecting these colors from a predefined series of palettes.

Using palettes

The two most significant colors we can edit in a material theme are the primary colors. They are applied directly to the status and app bars and give an app a distinctive look without affecting the uniform feel of the platform as a whole. Both these colors should be selected from the same color palette. There are many such palettes available, and the entire collection can be found at www.google.com/design/spec/style/color.html#color-color-palette.

Whichever palette you decide to use as your primary colors, Google recommend that you use shades with values of **500** and **700**:

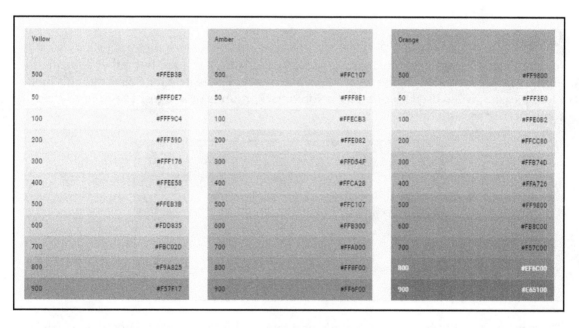

This does not have to be enforced too strictly, but it usually a good idea to stick close to these values and to always select two shades of the same color.

The theme editor can be very helpful here; not only do its solid color blocks offer tooltips telling us the shade value, but once we have picked a primary color, it will suggest a suitable darker version.

Our choice of primary shades needs to be considered when selecting the accent color. This will be applied to switches and highlights and needs to contrast nicely with the primary color. There are no simple rules governing which colors contrast with each other other than pick a color that looks good and has a light shade value of **100**, or close by.

It is possible to change the color of the navigation bar at the foot of the screen with `navigationBarColor`, but this is not recommended as the navigation bar should not really be thought of as part of your app.

Most of the other theme settings can be left as they are for most purposes. However, if you wish to change text colors there are one or two things to note.

Customizing text

Material text does not generate lighter shades by using lighter hues, but rather by using the alpha channel to create varying levels of **transparency**. The reason for this is that this looks more pleasing when used on different colored backgrounds or images. The rules for text transparency are as follows:

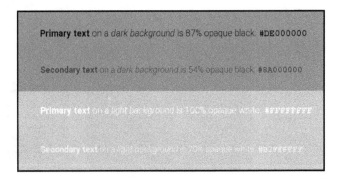

There is a lot that can be done with styles and themes, but for now it is enough to create a simple color scheme and know that it will be applied consistently across an application. Our next task will be to see how we can expand the sandwich ingredient objects we considered earlier into a user-friendly interface. No doubt one of the best ways to entice a user is with an appetizing photograph.

Adding image resources

One of the most interesting challenges that Android offers is the vast number of screen densities and sizes that we have to cater for. This is particularly true when it comes displaying bitmap images, where there are two competing issues that need to be resolved:

- Low resolution images display very poorly when stretched to fit on large or high resolution screens
- High quality images use up far more memory than is needed when displayed on smaller, low density screens.

Screen sizes aside, the problem of differing screen densities is mostly solved with the use of **density-independent pixels (dp)**.

Managing screen densities

Dps are an abstract unit of measurement based on a screen displaying 160 dpi. This means that a widget with a width of 320 dp will always be 2" wide regardless of screen density. When it comes to the actual physical dimensions of a screen, this can be managed with a variety of layout types, libraries, and properties such as weight and gravity, but for now we will look at how to provide images that suit the widest possible range of screen densities.

The Android system divides screen densities with the following qualifiers:

- Low density (ldpi) – **120 dpi**
- Medium density (mdpi) – **160 dpi**
- High density (hdpi) – **240 dpi**
- Extra-high density (xhdpi) – **320 dpi**
- Extra-extra-high density (xxhdpi) – **480 dpi**
- Extra-extra-extra-high density (xxxhdpi) – **640 dpi**

 During an app's installation, each device will only download images that match its own specifications. This saves memory on older devices and still provides the richest possible visual experience on devices that are capable.

From a developer's point of view, it might seem that we have to generate six versions of every image we want to include in any given project. Thankfully, this is not usually the case. The difference between a 640 dpi image and, say, a 320 dpi image is hardly noticeable on most handheld devices, and considering that most users of our sandwich builder app will simply want to scroll through a menu of ingredients, rather than scrutinize the quality of our imagery, we can safely provide images for medium, high, and extra-high density devices only.

 A good rule of thumb when considering image quality for high-end devices is to compare our image sizes with those produced by the device's native camera. It is unlikely that providing larger images will improve user experience enough to justify the extra memory required.

In our example here, we will want to provide images that fit onto a card view that will occupy most of the screen width in portrait mode. For now, find an image that is roughly 2,000 pixels in width. In the following example, it is called `sandwich.png` and is 1,920 by 1,080 pixels in size. Your image does not have to match these dimensions, but later we will see how well-selected image proportions are considered a significant part of good UI practice.

An image that has a width of 1,920 pixels would be six inches wide when displayed on an extra-high density device displaying 320 dpi. We will assume, for now at least, that our app will be accessed from mobile devices, rather than computers or televisions, so even on a high density, 10" tablet, six inches will be more than enough for our purposes. Next, we will see how to prepare for other screen densities too.

Using designated resources

Providing alternative bitmaps to suit a variety of screen densities is easily achieved by assigning **designated resource directories** to contain images configured for specific screen densities. From Android Studio, we can create such directories from the project explorer with these steps:

1. First, create a `New | Directory` from the `res` folder and call it `drawable-mdpi`.
2. Next, create two more sibling directories called `drawable-hdpi` and `drawable-xhdpi`.

3. Open these new folders directly by selecting **Show in Explorer** in the `drawable` context menu from the project explorer.

4. Add the `sandwich.png` image to the `drawable-xhdpi` folder.

5. Make two copies of this image and re-scale them so that one is 3:4 the original scale and the other 1:2 scale.

6. Place these copies in the `drawable-hdpi` and `drawable-mdpi` directories respectively.

These variations will now appear in the project explorer, as seen here:

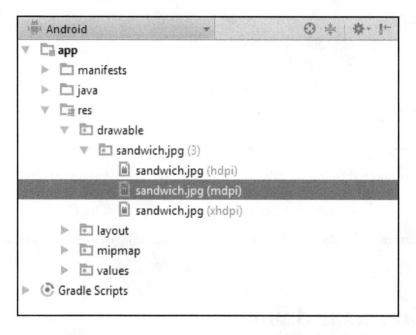

We can now rest assured that only the most suitable and memory efficient image resource will be downloaded according to a device's native screen density. To see how this looks, add the following image view to the project's `activity_main.xml` file:

```xml
<ImageView
    android:layout_width="wrap_content"
    android:layout_height="wrap_content"
    android:src="@drawable/sandwich" />
```

The output can be viewed with the preview screens on any emulator or real device:

The nice thing about this approach is that once we have the variations of our image designated correctly, we can simply refer to it as `@drawable/sandwich` and forget about the actual device it is being viewed on or which directory it is stored in.

This leaves us free to explore how we might include our images as part of a broader interface.

Creating a card view

The card view is one of the most recognizable material design components. It is designed to show several pieces of content that all apply to a single subject. This content is usually a combination of graphics, text, action buttons, and icons, and cards are a great way to present a selection of choices in a uniform way. This makes it a good choice for displaying our sandwich ingredients and related information such as price or calorific value. We will use the factory pattern from the previous chapter to do this, but before we see what code needs changing, let's take a look at how we implement the card view in the first place.

Understanding card view properties

If your minimum target SDK is 21 or greater, then the **CardView** widget will be included as standard. Otherwise, you will need to include the cardview support library. This is easily added in the `build.gradle` file by including the following highlighted line:

```
dependencies {
    compile fileTree(dir: 'libs', include: ['*.jar'])
    testCompile 'junit:junit:4.12'
    compile 'com.android.support:appcompat-v7:23.4.0'
    compile 'com.android.support:cardview-v7:23.4.0'
}
```

As the name of the support library suggests, we can only support card views as far back as API level 7.

It is not necessary to edit the `build.gradle` file manually, although it is useful to know how, as it can be done more simply via the `File | Project Structure...` menu and selecting the items shown here:

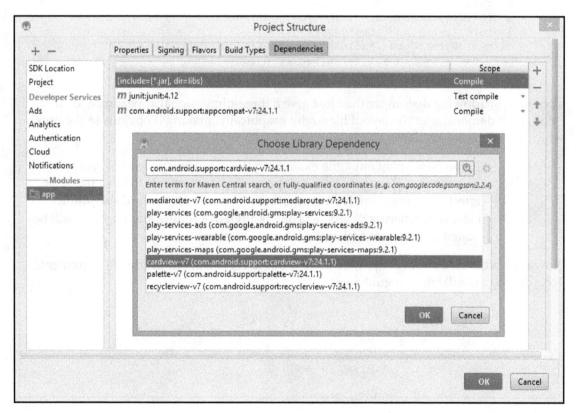

Some developers version their support libraries with a + symbol, like so: `com.android.support:cardview-v7:23.+`. This is an attempt to anticipate future libraries. This generally works very well, but it does not guarantee that these apps might not crash at a later date. It is a little more time-consuming, but far smarter, to use the compiled SDK version during development and then update the app regularly once it is published.

You will need to rebuild the project before we can add the card view to our layout. First, we will need to set some of the card's properties. Open the `res/values/dimens.xml` file and add the following three new dimension resources:

```xml
<dimen name="card_height">200dp</dimen>
<dimen name="card_corner_radius">4dp</dimen>
<dimen name="card_elevation">2dp</dimen>
```

Now we can add the card as a widget in the main XML activity file, like so:

```xml
<android.support.v7.widget.CardView
xmlns:card_view="http://schemas.android.com/apk/res-auto"
    android:layout_width="match_parent"
    android:layout_height="@dimen/card_height"
    android:layout_gravity="center"
    card_view:cardCornerRadius="@dimen/card_corner_radius"
    card_view:cardElevation="@dimen/card_elevation">
</android.support.v7.widget.CardView>
```

The use of shadowing does more than just give a three-dimensional appearance to an interface; it demonstrates the layout hierarchy graphically, making it obvious to the user which functions are available.

If you have spent any time examining the card view properties, you will have noticed the `translationZ` property. This appears to have the same effect as `elevation`. However, `elevation` will set the card's absolute elevation, whereas `translationZ` is a relative setting and its value will be added or subtracted from the current elevation.

Now we have a card view set up, we can fill it out to represent our sandwich ingredients according to material design guidelines.

Applying CardView metrics

Design guidelines are very clear about such issues as typeface, padding, and scale. Once we start using the CoordinatorLayout, a lot of these settings will be set automatically, but for now, it is a good idea to see how these metrics are applied.

There are many different patterns for cards, and a full description of them can be found here:

`www.google.com/design/spec/components/cards.html`

The one we will create here will contain an image, three text items, and an action button. Cards can be considered as container objects, and as such, normally contain their own root layout. This can be placed directly inside the card view itself, but it makes for more readable and more flexible code if we create the card content as a separate XML layout.

We will need at least one image for this next exercise. According to material design, photographic images should be clear, bright, simple, and present a single, unambiguous subject. For example, if we wanted to add coffee to our menu, the image on the left would be the most suitable of the two:

Card images need to have a width to height ratio of 16:9 or 1:1. Here, we will use 16:9, and ideally we should produce scaled versions to suit various screen densities, but as this is only a demonstration, we can be lazy and just place the originals directly into the `drawable` folder. This approach is far from best practice, but is fine for preliminary testing.

Once you have sourced and saved your images, the next step is to create a layout for our card:

1. From the project explorer, navigate to `New | XML | Layout XML File` and call it `card_content.xml`. Its root view group should be a linear layout with a vertical orientation, and it should look like this:

```
<LinearLayout xmlns:android="http://schemas.android.com/apk/res/android"
    android:id="@+id/card_content"
    android:layout_width="match_parent"
    android:layout_height="match_parent"
    android:orientation="vertical">
</LinearLayout>
```

2. Using either the graphical or text editor, create a layout structure to match the **Component Tree** seen here:

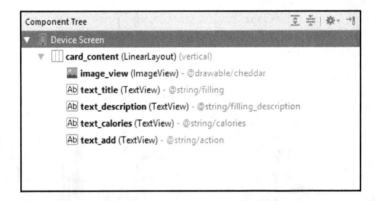

3. Now include this layout inside the card view from the main activity layout file, like this:

```
<android.support.v7.widget.CardView
    android:id="@+id/card_view"
    android:layout_width="match_parent"
    android:layout_height="wrap_content">

    <include
        android:id="@+id/card_content"
        layout="@layout/card_content" />

</android.support.v7.widget.CardView>
```

Although it is editable, the recommended elevation for a card view is 2 dp, unless it has been selected and/or is being moved, in which case it has an elevation of 8 dp.

As you will no doubt know, the use of strings that are hard-coded into XML resources are strongly discouraged. If nothing else, this makes the process of translating our apps into other languages almost impossible. However, during the early stages of layout design, it helps to provide some placeholder values to give an idea of how a layout might look. Later, we will control card content using Java, and select this content based on user input; but for now, we will select some typical values so that we can see the effect our settings have easily and quickly. To see how this is of use, add the following properties, or equivalent, to the `strings.xml` file in the `values` directory:

```
<string name="filling">Cheddar Cheese</string>
<string name="filling_description">A ripe and creamy cheddar from the south
west</string>
<string name="calories">237 kcal per slice</string>
<string name="action">ADD</string>
<string name="alternative_text">A picture of some cheddar cheese</string>
```

For now, we will use these placeholders to evaluate any changes we make as we make them. The layouts we just created should, when viewed as a preview, look something like this:

Converting this into a materially designed component requires nothing more than some formatting and a little knowledge of material guidelines.

The metrics for this layout are as follows:

- The image must have a ratio of 16:9
- The title text should be 24 sp
- The descriptive text is 16 sp
- The margins around the bottom right and left of the text is 16 dp
- The margin above the title text is 24 dp
- The action text as 24 sp and takes its color from the accent

These properties are very easily set from either the properties panel or by editing the XML directly. There are one or two things not mentioned here, so it is worth taking a look at each element separately.

First, it is essential to point out that these values should never be described literally in the code as they are in the following snippet; for example, `android:paddingStart="24dp"` should be coded as something like `android:paddingStart="@dimen/text_paddingStart"` with `text_paddingStart` being defined in the `dimens.xml` files. Here, the values have been hard-coded only to simplify the explanation.

The code for image view at the top should look like this:

```
<ImageView
        android:id="@+id/image_view"
        android:layout_width="match_parent"
        android:layout_height="wrap_content"
        android:contentDescription="@string/alternative_text"
        android:src="@drawable/cheddar" />
```

This is very straightforward, but do note the use of `contentDescription`; this is used when a visually impaired user has set accessibility options so that images can be appreciated by having their descriptions read out by the device's voice synthesizer.

Beneath this are the following three text views.

```
<TextView
    android:id="@+id/text_title"
    android:layout_width="wrap_content"
    android:layout_height="wrap_content"
    android:paddingEnd="24dp"
    android:paddingStart="24dp"
    android:paddingTop="24dp"
    android:text="@string/filling"
    android:textAppearance="?android:attr/textAppearanceLarge"
```

```
        android:textSize="24sp" />

    <TextView
        android:id="@+id/text_description"
        android:layout_width="wrap_content"
        android:layout_height="wrap_content"
        android:paddingEnd="24dp"
        android:paddingStart="24dp"
        android:text="@string/filling_description"
        android:textAppearance="?android:attr/textAppearanceMedium"
        android:textSize="14sp" />

    <TextView
        android:id="@+id/text_calories"
        android:layout_width="wrap_content"
        android:layout_height="wrap_content"
        android:layout_gravity="end"
        android:paddingBottom="8dp"
        android:paddingStart="16dp"
        android:paddingEnd="16dp"
        android:paddingTop="16dp"
        android:text="@string/calories"
        android:textAppearance="?android:attr/textAppearanceMedium"
        android:textSize="14sp" />
```

These too are very simple to follow. All that really needs pointing out is the use of `Start` and `End` as opposed to `Left` and `Right` to define padding and gravity, as this helps apply our layout correct itself when translated into languages where text runs from right to left. We also included the `textAppearance` property, which may appear redundant as we also set text size directly. Attributes such as `textAppearanceMedium` are useful as not only do they automatically apply text coloring according to our customized theme, they will also adjust their size according to individual users' global text size settings.

This leaves only the action button at the bottom, and as this uses a text view rather than a button, this may require a little explanation. The XML looks like this:

```
    <TextView
        android:id="@+id/text_add"
        android:layout_width="wrap_content"
        android:layout_height="wrap_content"
        android:layout_gravity="end"
        android:clickable="true"
        android:paddingBottom="16dp"
        android:paddingEnd="40dp"
        android:paddingLeft="16dp"
        android:paddingRight="40dp"
        android:paddingStart="16dp"
```

```
android:paddingTop="16dp"
android:text="@string/action"
android:textAppearance="?android:attr/textAppearanceLarge"
android:textColor="@color/colorAccent"
android:textSize="24sp" />
```

There are two reasons why we chose a text view here where it would seem to make sense to use a button widget. Firstly, Android recommends the use of a **flat button** where only the text is visible on card views; and secondly, the touchable area that triggers an action needs to be larger than the text itself. This is easily performed by setting the padding properties as we have previously. To make a text view behave like a button, we only need to add the line `android:clickable="true"`.

Our finished card should now look something like this:

There is a lot more to the design of card views, but this should serve as a good introduction to some of the principles we need to follow, and for now, we can see how these new ways of presenting our objects reflect on our factory pattern code.

Updating the factory pattern

One of the beauties of design patterns is the ease by which they can adapted to suit any changes we wish to make. We could, if we chose, leave our factory code as it is and use the single string output to direct the client code to the appropriate dataset. It is more in keeping with the nature of patterns, though, to adapt them to match our slightly more complex ingredient objects.

The thought that went into our code structure in the last chapter now pays off, as although we need to edit our interfaces and concrete examples, we can leave the factory classes themselves just as they are, and this demonstrates one of the advantages of patterns rather nicely.

Using the four criteria we used to build our card, our updated interfaces could look like this:

```java
public interface Bread {

    String image();
    String name();
    String description();
    int calories();
}
```

Individual objects could look like so:

```java
public class Baguette implements Bread {

    @Override
    public String image() {
        return "R.drawable.baguette";
    }

    @Override
    public String name() {
        return "Baguette";
    }

    @Override
    public String description() {
        return "Fresh and crunchy";
    }

    @Override
    public int calories() {
        return 150;
    }

}
```

As we move forward, we will need more properties for our objects, such as price and whether they are vegetarian or contain nuts, and as our objects become more complex we will have to apply more sophisticated ways to manage our data, but in principle there is nothing wrong with the approach we are using here. It may be bulky, but it is certainly easy to read and maintain. Factory patterns are clearly very useful, but they only create single objects. To model a sandwich more realistically, we need to be able to put *ingredient* objects together and treat the entire collection as a single *sandwich* object. This is where the builder pattern comes in.

Applying a builder pattern

The builder design pattern is one of the most useful creational patterns as it builds larger objects from smaller ones. This is precisely what we want to do to construct a sandwich object from a list of ingredients. The builder pattern has a further advantage in that optional features are easy to include later. As before, we will begin by creating an interface; we will call it Ingredient and use it to represent both bread and filling. This time, we will need to represent calories as an integer, as we will need to calculate the total amount in a finished sandwich.

Open an Android Studio project, or start a new one, and follow the proceeding steps to create a basic sandwich builder pattern:

1. Create a new interface called Ingredient.java, and complete it like so:

```
public interface Ingredient {

    String name();
    int calories();
}
```

2. Now create an abstract class for Bread like this:

```
public abstract class Bread implements Ingredient {

    @Override
    public abstract String name();

    @Override
    public abstract int calories();
}
```

3. And create an identical one called `Filling`.

4. Next, create concrete classes of `Bread`, like this:

```
public class Bagel extends Bread {

    @Override
    public String name() {
        return "Bagel";
    }

    @Override
    public int calories() {
        return 250;
    }
}
```

5. Do the same for `Filling`. Two classes of each type should be enough for demonstration purposes:

```
public class SmokedSalmon extends Filling {

    @Override
    public String name() {
        return "Smoked salmon";
    }

    @Override
    public int calories() {
        return 400;
    }
}
```

6. Now we can create our `Sandwich` class:

```
public class Sandwich {
    private static final String DEBUG_TAG = "tag";

    // Create list to hold ingredients
    private List<Ingredient> ingredients = new ArrayList<Ingredient>();

    // Calculate total calories
    public void getCalories() {
        int c = 0;

        for (Ingredient i : ingredients) {
            c += i.calories();
        }
```

```
            Log.d(DEBUG_TAG, "Total calories : " + c + " kcal");
    }

    // Add ingredient
    public void addIngredient(Ingredient ingredient) {
        ingredients.add(ingredient);
    }

    // Output ingredients
    public void getSandwich() {

        for (Ingredient i : ingredients) {
            Log.d(DEBUG_TAG, i.name() + " : " + i.calories() + " kcal");
        }
    }

}
```

7. Finally, create the `SandwichBuilder` class like so:

```
public class SandwichBuilder {

    // Off the shelf sandwich
    public static Sandwich readyMade() {
        Sandwich sandwich = new Sandwich();

        sandwich.addIngredient(new Bagel());
        sandwich.addIngredient(new SmokedSalmon());
        sandwich.addIngredient(new CreamCheese());

        return sandwich;
    }

    // Customized sandwich
    public static Sandwich build(Sandwich s, Ingredient i) {

        s.addIngredient(i);
        return s;
    }
}
```

This completes our builder design pattern, for now at least. When viewed as a diagram, it looks like this:

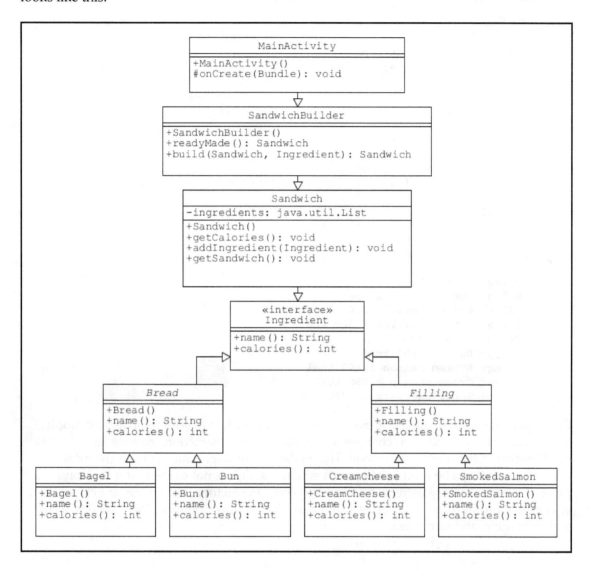

Here, we have provided the builder with two functions: to return a ready-made sandwich and a user-customized one. We have no working interface as yet, but we can simulate user choice via our client code.

We have also delegated output responsibilities to the `Sandwich` class itself, and this is often a good idea as it helps keep our client code clean and obvious, as you can see here:

```
// Build a customized sandwich
SandwichBuilder builder = new SandwichBuilder();
Sandwich custom = new Sandwich();
// Simulate user selections
custom = builder.build(custom, new Bun());
custom = builder.build(custom, new CreamCheese());
Log.d(DEBUG_TAG, "CUSTOMIZED");
custom.getSandwich();
custom.getCalories();

// Build a ready made sandwich
Sandwich offTheShelf = SandwichBuilder.readyMade();
Log.d(DEBUG_TAG, "READY MADE");
offTheShelf.getSandwich();
offTheShelf.getCalories();
```

This should produce an output along these lines:

```
D/tag: CUSTOMIZED
D/tag: Bun : 150 kcal
D/tag: Cream cheese : 350 kcal
D/tag: Total calories : 500 kcal
D/tag: READY MADE
D/tag: Bagel : 250 kcal
D/tag: Smoked salmon : 400 kcal
D/tag: Cream cheese : 350 kcal
D/tag: Total calories : 1000 kcal
```

One of the biggest advantages of the builder is how easy it is to add, remove, and modify the concrete classes, and even changes in the interface or abstractions require no modification of the client source code. This makes the builder pattern one of the most powerful, and it can be used in numerous situations. This is not to say that it is always preferable to the factory pattern. For simple objects, the factory is often the best choice, and of course, patterns exist on different scales and it is not uncommon to find factories nested within builders and vice versa.

Summary

In this chapter, we covered a lot of material about how we present our products, and this is a vital element to any successful app. We saw how to manage color and text schemes and then went on to the more serious issue of managing the wide number of screen densities our apps may find themselves running on.

One of material design's most frequently used components, the card view, was covered next, and the importance of support libraries was emphasized, and in particular the design library. We will need to look further into this library, as it is vital for creating the kind of layouts and interaction our apps deserve. The next chapter will concentrate on more of these visual elements, focusing on the more commonly found material components such as app-bars and sliding drawers.

3
Material Patterns

So far in this book we have looked at how to represent objects and collections of objects by using design patterns to create them and the card view to display them. Before we can start to put together a working application, we need to consider how a user will input their selections. There are numerous ways to gather information from the user on a mobile device, such as menus, buttons, icons, and dialogs. Android layouts generally have an application bar (previously known as the action bar) that usually sits at the top of the screen just under the status bar and layouts that implement material design, very often employ a sliding navigation drawer to provide access to an app's top-level functions.

As is often the case, the use of the support libraries, and in particular the **design library**, makes implementing material patterns such as the navigation bar remarkably easy, and material design contains visual patterns of its own that help encourage best UI practices. In this chapter, we will see how to implement the **app-bar**, the **navigation view**, and explore some of the visual patterns that material design provides. We will conclude with a quick look at the **singleton pattern**.

In this chapter, you will learn how to do the following:

- Replace the action bar with an app-bar
- Add action icons with the Asset Studio
- Apply app-bar actions
- Control the app-bar at runtime
- Use a drawer layout
- Add menus and sub-menus
- Apply ratio keylines
- Include a drawer listener
- Add fragments to an app
- Manage the fragment back stack

The app-bar

Android applications have always contained a toolbar at the top of the screen. This was traditionally used to provide a title along with access to a top-level menu, and was called the action bar. Since Android 5 (API level 21) and the inception of material design, it has been possible to replace this with the far more flexible app-bar. The app bar allows us to set its color, place it anywhere on the screen, and include a wider range of content than its predecessor.

Most Android Studio templates use themes that include the old action bar as default, and the first thing we will need to do is remove the old version. To see how we remove the old action bar and replace it with a customized app-bar, follow these steps:

1. Start a new Android project using an empty activity template and set your material theme using the theme editor.
2. Open the `styles.xml` file and edit the `style` definition to match the one here:

```xml
<style name="AppTheme" parent="Theme.AppCompat.Light.NoActionBar">
```

3. Create a new XML file alongside `activity_main.xml` and call it `toolbar.xml`.
4. Complete it like so:

```xml
<android.support.v7.widget.Toolbar
    xmlns:android="http://schemas.android.com/apk/res/android"
    android:id="@+id/toolbar"
    android:layout_width="match_parent"
    android:layout_height="?attr/actionBarSize"
    android:background="?attr/colorPrimary"
    android:theme="@android:style/Theme.Material"
    android:translationZ="4dp" />
```

5. Next, add the following element to the `activity_main.xml` file:

```xml
<include
    android:id="@+id/toolbar"
    layout="@layout/toolbar" />
```

6. Finally, edit the margin values in the `dimens.xml` file as seen here:

```xml
<resources>
    <dimen name="activity_horizontal_margin">0dp</dimen>
    <dimen name="activity_vertical_margin">0dp</dimen>
</resources>
```

This toolbar is like any other ViewGroup in that it sits inside the root layout, so unlike the original action bar, it is not flush against the edges of the screen. This is why we needed to adjust the layout margins. Later, we will employ the CoordinatorLayout, which will automate much of this for us, but for now it is useful to see how it all works.

The toolbar is now positioned and shaded like the original but has none of the content or functions. This can be done in the Java element of the activity by editing the onCreate() method like this:

```
@Override
protected void onCreate(Bundle savedInstanceState) {
    super.onCreate(savedInstanceState);
    setContentView(R.layout.activity_main);

    Toolbar toolbar = (Toolbar) findViewById(R.id.toolbar);
    if (toolbar != null) {
        setSupportActionBar(toolbar);
    }
}
```

This will generate an error. This is because there are two possible libraries that could be imported here. Press Alt + Enter and select the support version of the Toolbar like so:

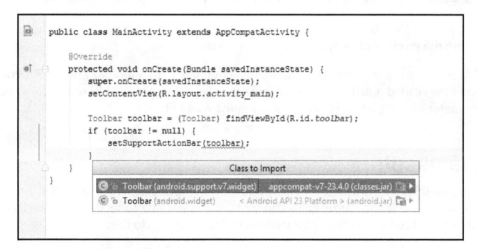

To save time when working with Java, change the settings so that Java libraries are automatically imported when included in code. This is done from the **File** | **Settings** menu with **Editor** | **General** | **Auto Import**.

Testing the project on an emulator running API 20 or lower will immediately demonstrate one of the shortfalls of the AppCompat theme; despite declaring a color for our status bar with `colorPrimaryDark`, which works perfectly on API 21 and higher, here it is still black:

However, this and the absence of natural-looking shadows is a small price to pay considering the number of people we can now reach.

Now that we have replaced the old-fashioned action bar with a toolbar and set it as the app bar (sometimes called a primary toolbar) we can take a closer look at how it works and how to apply material-compliant action icons using the Asset Studio.

Image assets

It is quite possible to include text menus in the app-bar, but due to the limited space, it is more normal to use icons. Android Studio provides access to a collection of material icons via its Asset Studio. The following steps demonstrate how to do this:

1. From the drawable folder's menu in the project explorer, select **New | Image Asset**.

2. Then select **Action Bar and Tab Icons** as the **Asset Type** and then click on the **Clipart** icon and select an icon from the collection of clip art:

3. This image must be trimmed, with 0% padding.
4. Choose a theme depending on whether your toolbar background color is light or dark.

5. Provide a suitable name and click on **Next**:

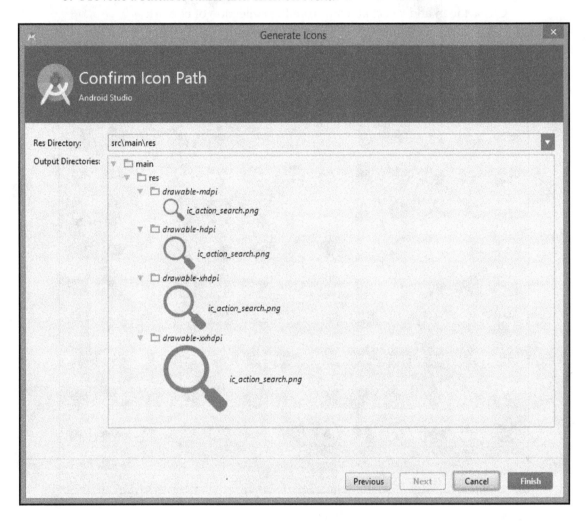

 A much larger collection of material icons can be downloaded from the following URL: `https://design.google.com/icons`

The asset studio automatically creates icons for us across four screen densities and places them in the correct folders so that they are deployed on the appropriate devices. It even applies the required **54% opaque black** that material design uses for icons. All we have to do to include these in our app bar is to add an icon property to the appropriate menu item. Later, we will use a navigation drawer to provide top-level access, but to see how to use an app-bar, we will add a search function. The icon we chose for this is called `ic_action_search`.

Applying actions

Action icons are kept in the drawable folders and can be included in our action bar by including `items` within menu XML files. Depending on which template you used when first creating the project, you may have to add a new directory, `res/menu`, and a file called `main.xml` or `menu_main.xml` or whatever you choose as a **New | Menu resource file**. Actions can be added like so:

```xml
<menu xmlns:android="http://schemas.android.com/apk/res/android"
    xmlns:app="http://schemas.android.com/apk/res-auto"
    xmlns:tools="http://schemas.android.com/tools"
    tools:context="com.example.kyle.appbar.MainActivity">

    <item
        android:id="@+id/action_settings"
        android:orderInCategory="100"
        android:
        app:showAsAction="collapseActionView" />

    <item
        android:id="@+id/action_search"
        android:icon="@drawable/ic_action"
        android:orderInCategory="100"
        android:
        app:showAsAction="ifRoom" />
</menu>
```

Note that the preceding example uses a reference to a string resource, and so must be accompanied by a definition in the `strings.xml` file like so:

```xml
<string name="menu_search">Search</string>
```

Menu items are automatically included in the app bar, with the title being taken from the `string name="app_name"` definition in the strings file. When constructed in this fashion, these components are positioned according to material guidelines.

To see this in action, follow these steps:

1. Open the main Java activity and add this field:

```
private Toolbar toolbar;
```

2. Then add these lines to the onCreate() method:

```
Toolbar toolbar = (Toolbar) findViewById(R.id.toolbar);
    if (toolbar != null) {
        setSupportActionBar(toolbar);
    }

toolbar = (Toolbar) findViewById(R.id.toolbar);
toolbar.setTitle("A toolbar");
toolbar.setSubtitle("with a subtitle");
```

3. Lastly, add the following method to the activity:

```
@Override
public boolean onCreateOptionsMenu(Menu menu) {
    MenuInflater inflater = getMenuInflater();
    inflater.inflate(R.menu.menu_main, menu);
    return true;
}
```

We should now be able to see our new toolbar on a device or emulator:

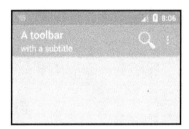

Being able to add any view we like to a toolbar makes it far more effective than the old action bar. We can have more than one at a time and they can even be placed elsewhere by applying layout gravity properties. The toolbar even comes with its own methods as we saw previously with the title and subtitle. We can also add icons and logos with these methods, but before we do so, it would be a good idea to explore app bar best practice according to material design guidelines.

App bar structure

Although the techniques we have applied here conform to material guidelines without us having to do very much other than ensure its height, there will still be times when we are replacing the action bar with a custom toolbar layout, and we will need to know how to space and position the components. These are slightly different for tablets and desktops.

Phones

There are just a few simple structural rules to remember when it comes to app bars. These cover margins, padding, width, height, and positioning, and they differ across platforms and screen orientation:

- The `layout_height` of an app bar in portrait mode is 56 dp and 48 dp in landscape.
- App bars fill either the screen width or the width of their containing column. They cannot be divided into two. They have a `layout_width` of `match_parent`.
- An app bar has an `elevation` 2 dp greater than the sheet of material it controls.
- The exception to the preceding rule is if a card or dialog has its own toolbar, then the two can share the same elevation.
- App bars have padding of exactly 16 dp. This means the contained icons must have no padding or margins of their own and therefore share edges with this margin:

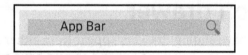

- The title text takes its color from your theme's primary text color and icons from secondary text.
- The title should be positioned 72 dp from the left of the toolbar and 20 dp from the bottom. This applies even when the toolbar is expanded:

- The Title's text size is set with
  `android:textAppearance="?android:attr/textAppearanceLarge"`.

Tablets

When constructing app bars for tablets and desktops, the rules are identical, with the following exceptions:

- The toolbar height is always `64 dp`.
- The title is indented by `80 dp` and does not move down when the bar is expanded.
- The app bar's padding is `24 dp`, with the exception of the top, where it is `20 dp`.

We have succeeded in constructing an app bar according to material guidelines, but action icons are of no use if they do not perform an action. In essence, when an app-bar assumes action bar functionality, it is really simply an access point to a menu. We will return to menus and dialogs later, but for now we will take a quick look at how toolbars can be manipulated at runtime with Java code.

The changes that have been made to the old action bar make it an easy and intuitive view to place global actions. The space, however, is limited, and for a more complex and graphical navigation component, we can turn to the sliding drawer.

The navigation drawer

Although it is possible to have sliding drawers appear from either side of the screen, the navigation drawer should always be on the left and should have a higher elevation than all other views apart from the status and navigation bars. Think of the navigation drawer as a permanent fixture that spends most of its time hidden just off the edge of the screen:

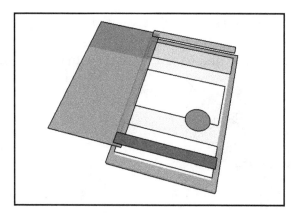

Prior to the design library, components such as the navigation view had to be constructed from other views, and although the library vastly simplifies this process and saves us having to implement many material principles by hand, there are still several guidelines that we need to be aware of. The best way to appreciate these is by building a navigation sliding drawer from scratch. This will involve creating the layouts, applying material guidelines regarding component ratios, and connecting all this together with code.

Drawer construction

You will no doubt have noticed when setting up projects that Android Studio provides a **Navigation Drawer Activity** template. This creates much of the structure we might need and saves quite a bit of work. Once we have decided what features our sandwich building app will have, we will use this template. However, it is far more instructive to put one together from scratch to see how it works, and with this in mind, we will create a drawer layout that requires icons that can easily be found via the Asset Studio:

1. Open an Android Studio project with a minimum SDK level of 21 or higher and provide it with your own customized colors and theme.
2. Add the following line to your `styles.xml` file:

```
<item name="android:statusBarColor">
@android:color/transparent
</item>
```

3. Make sure you have the following dependency compiled:

```
compile 'com.android.support:design:23.4.0'
```

4. If you are not using the same project we used in the previous section, set up an app-bar layout called `toolbar.xml`.

5. Open `activity_main` and replace the code with the following:

```xml
<android.support.v4.widget.DrawerLayout
xmlns:android="http://schemas.android.com/apk/res/android"
    xmlns:app="http://schemas.android.com/apk/res-auto"
    xmlns:tools="http://schemas.android.com/tools"
    android:id="@+id/drawer"
    android:layout_width="match_parent"
    android:layout_height="match_parent"
    android:fitsSystemWindows="true"
    tools:context=".MainActivity">

    <LinearLayout
        android:layout_width="match_parent"
        android:layout_height="match_parent"
        android:orientation="vertical">

        <include
            android:id="@+id/toolbar"
            layout="@layout/toolbar" />

        <FrameLayout
            android:id="@+id/fragment"
            android:layout_width="match_parent"
            android:layout_height="match_parent">
        </FrameLayout>

    </LinearLayout>

    <android.support.design.widget.NavigationView
        android:id="@+id/navigation_view"
        android:layout_width="wrap_content"
        android:layout_height="match_parent"
        android:layout_gravity="start"
        app:headerLayout="@layout/header"
        app:menu="@menu/menu_drawer" />

</android.support.v4.widget.DrawerLayout>
```

As you can see, the root layout here is the **DrawerLayout** as provided by the support library. Note the `fitsSystemWindows` property; this is what makes the drawer extend up to the top of the screen under the status bar. Having set the `statusBarColor` to `android:color/transparent` in the style, the drawer is now visible through the status bar.

This effect is not available on devices running Android versions older than 5.0 (API 21), even with AppCompat, and this will alter the apparent aspect ratio of the header and clip any images. To counter this, create an alternative `styles.xml` resource that does not set the `fitsSystemWindows` property.

The rest of the layout consists of a LinearLayout and the **NavigationView** itself. The linear layout contains our app bar and an empty **FrameLayout**. FrameLayouts are the simplest of layouts, containing only a single item and generally used as a placeholder, which in this case will contain content based on the user's selection from the navigation menu.

As can be seen from the preceding code, we will need a layout file for the header and a menu file for the drawer itself. The `header.xml` file should be created in the `layout` directory and look like this:

```xml
<?xml version="1.0" encoding="utf-8"?>
<RelativeLayout xmlns:android="http://schemas.android.com/apk/res/android"
    android:layout_width="match_parent"
    android:layout_height="header_height"
    android:background="@drawable/header_background"
    android:orientation="vertical">

    <TextView
        android:id="@+id/feature"
        android:layout_width="wrap_content"
        android:layout_height="wrap_content"
        android:layout_above="@+id/details"
        android:gravity="left"
        android:paddingBottom="8dp"
        android:paddingLeft="16dp"
        android:text="@string/feature"
        android:textColor="#FFFFFF"
        android:textSize="14sp"
        android:textStyle="bold" />

    <TextView
        android:id="@+id/details"
        android:layout_width="wrap_content"
        android:layout_height="wrap_content"
        android:layout_alignStart="@+id/feature"
        android:layout_alignParentBottom="true"
        android:layout_marginBottom="16dp"
        android:gravity="left"
        android:paddingLeft="16dp"
        android:text="@string/details"
        android:textColor="#FFFFFF"
        android:textSize="14sp" />
```

```
    </RelativeLayout>
```

You will need to add the following value to the `dimens.xml` file:

```
<dimen name="header_height">192dp</dimen>
```

As you will see, we will need an image for the header. Here, it is called `header_background` and should have an aspect ratio of 4:3.

If you test this layout on devices with different screen densities, you will very soon see that this aspect ratio is not maintained. This can be easily countered in a similar way to the manner that we manage image resources, by using configuration qualifiers. To do this, follow the simple steps outlined here:

1. Create new directories for each density range with names such as `values-ldpi`, `values-mdpi`, and so on up to `values-xxxhdpi`.
2. Make a copy of the `dimens.xml` file in each folder.
3. Set the value of `header_height` in each file to match that screen density.

The menu file is called `menu_drawer.xml` and should be placed in the `menu` directory, which you may need to create as well. Each item has an associated icon, and these can all be found in the Asset Studio. The code itself should match the following:

```
<?xml version="1.0" encoding="utf-8"?>
<menu xmlns:android="http://schemas.android.com/apk/res/android">

    <item
        android:id="@+id/drama"
        android:icon="@drawable/drama"
        android: />

    <item
        android:id="@+id/film"
        android:icon="@drawable/film"
        android: />

    <item
        android:id="@+id/sport"
        android:icon="@drawable/sport"
        android: />

    <item
        android:id="@+id/news"
        android:>
        <menu>
            <item
```

```
        android:id="@+id/national"
        android:icon="@drawable/news"
        android: />

    <item
        android:id="@+id/international"
        android:icon="@drawable/international"
        android: />

</menu>
    </item>
</menu>
```

Most of the metrics of sliding drawers and navigation views such as margins and text sizes are taken care of for us thanks to the design library. However, the size, position, and color of text on a drawer header are not. Despite sharing a background, the text should be thought of as a 56-dp high component in its own right. It should have an internal padding of 16-dp and an 8-dp spacing between the lines. This, along with the correct text color, size, and weight can be derived from the preceding code.

Ratio keylines

When an element such as a sliding drawer fills the entire height of a screen and is divided into vertical segments, as our drawer is, between header and content, then these divisions can occur only at certain points known as ratio keylines. These points are determined by the ratio between the width of the element and how far from the top the division occurs. There are six such ratios allowed in material layouts, and they are defined as width to height (width:height) and are as follows:

- 16:9
- 3:2
- 4:3
- 1:1
- 3:4
- 2:3

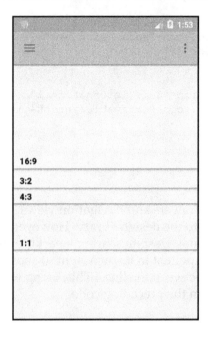

In the example here, a 4:3 ratio was chosen, and the width of the drawer is 256 dp. We could also have produced a header with a 16:9 ratio and set the `layout_height` at 144 dp.

Ratio keylines only relate to the distance from the top of the containing element; you cannot have one 16:9 view below another. However, you can place another view beneath this if it extends from the bottom of the top view down to another of the ratio keylines:

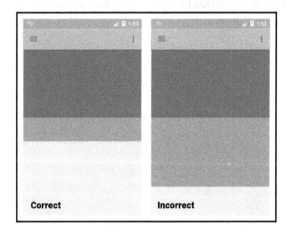

Activating the drawer

All that remains now is to implement some code in Java to get the layout working. This is done through a listener callback method that is called when the user interacts with the drawer. The following steps demonstrate how this is achieved:

1. Open the MainActivity file and add the following lines in the `onCreate()` method to replace the action bar with our toolbar:

```
toolbar = (Toolbar) findViewById(R.id.toolbar);
setSupportActionBar(toolbar);
```

2. Beneath this, add these lines to configure the drawer:

```
drawerLayout = (DrawerLayout) findViewById(R.id.drawer);
ActionBarDrawerToggle toggle = new ActionBarDrawerToggle(this,
drawerLayout, toolbar, R.string.openDrawer, R.string.closeDrawer) {


public void onDrawerOpened(View v) {
    super.onDrawerOpened(v);
}


public void onDrawerClosed(View v) {
    super.onDrawerClosed(v);
}

};

drawerLayout.setDrawerListener(toggle);
toggle.syncState();
```

3. Finally, add this code to set up the navigation view:

```
navigationView = (NavigationView) findViewById(R.id.navigation_view);

navigationView.setNavigationItemSelectedListener(new
NavigationView.OnNavigationItemSelectedListener() {

    @Override
    public boolean onNavigationItemSelected(MenuItem item) {

        drawerLayout.closeDrawers();

        switch (item.getItemId()) {
            case R.id.drama:
```

```
                Log.d(DEBUG_TAG, "drama");
                return true;
            case R.id.film:
                Log.d(DEBUG_TAG, "film");
                return true;
            case R.id.news:
                Log.d(DEBUG_TAG, "news");
                return true;
            case R.id.sport:
                Log.d(DEBUG_TAG, "sport");
                return true;
            default:
                return true;
            }
        }
    });
```

The preceding Java code allows us to view our drawer on a device or emulator but does very little when a navigation item is selected. What we really need to do is to actually be taken to another part of the app. This is very simply achieved, and we will come to it in a moment. First, there are one or two points in the preceding code that require a mention.

The line beginning `ActionBarDrawerToggle` is what causes the hamburger that opens the drawer to appear on the app bar, although you can of course open it with an inward swipe from the left of the screen. The two string arguments, `openDrawer` and `closeDrawer`, are for reasons of accessibility and are read out for users who are unable to see the screen clearly. They should say something like Navigation drawer opening and Navigation drawer closing. The two callback methods `onDrawerOpened()` and `onDrawerClosed()` were left empty here, but demonstrate where these events can be intercepted.

The call to `drawerLayout.closeDrawers()` is essential, as otherwise the drawer would remain open. Here, we used the debugger to test the output, but ideally what we want is for the menu to direct us to another part of the application. This is not a difficult task and also provides a good opportunity to introduce one of the SDK's most useful and versatile classes, the **fragment**.

Adding fragments

From what we have learned so far, it would be safe to imagine that separate activities would be used for apps with more than one function, and although this is often the case it can be an expensive drain on resources and activities always fill the entire screen. Fragments operate like mini-activities, in that they have both Java and XML definitions and many of the same callbacks and functionality that activities do. Unlike activities, fragments are not top level components and must reside within a host activity. The advantage of this is that we can have more than one fragment per screen.

To see how to do this, create a new Java class called something like `ContentFragment` and complete it as follows, making sure you import the `android.support.v4.app.Fragment` rather than the standard version:

```
public class ContentFragment extends Fragment {

    @Override
    public View onCreateView(LayoutInflater inflater, ViewGroup container,
Bundle savedInstanceState) {
        View v = inflater.inflate(R.layout.content, container, false);
        return v;
    }
}
```

As for the XML element, create a layout file called `content.xml` and place whatever views and widgets you choose inside. All that is needed now is the Java code to call it when a navigation item is selected.

Open the `MainActivity.Java` file and replace one of the Debug calls in the `switch` statement with this:

```
ContentFragment fragment = new ContentFragment();
android.support.v4.app.FragmentTransaction transaction =
getSupportFragmentManager().beginTransaction();
transaction.replace(R.id.fragment, fragment);
transaction.addToBackStack(null);
transaction.commit();
```

The example we have built here is solely to demonstrate the basic anatomy of drawer layouts and navigation views. Clearly, to add any real functionality, we would need a fragment for each item in our menu, and the line `transaction.addToBackStack(null);` is actually redundant unless we do so. Its function is to ensure that the order a user accesses each fragment is recorded by the system in the same way it records which activities are used, so that when they press the back key they will return to the previous fragment. Without it, they would be returned to the previous application and the container activity would be destroyed.

Right handed drawers

As a top-level navigation component, the sliding drawer should only ever slide in from the left and should follow the metrics outlined earlier. However, it is very easy to have drawers that slide in from the right, and for many secondary functions this can be desirable:

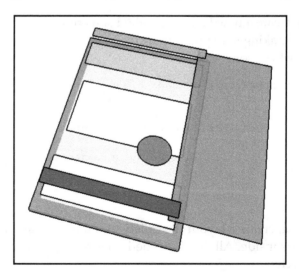

Making a sliding drawer appear from the right is simply a matter of setting layout gravity, for example:

```
android:layout_gravity="end"
```

Unlike the traditional navigation view, which should never be wider than the screen width minus the height of the primary app-bar, a right-handed drawer can extend across the entire screen.

The whole of this chapter has been about UI design, and we have not come across any design patterns. We could have used patterns here but chose to concentrate on the mechanics of Android UIs. We will see later in the book how useful the facade pattern can be to simplify coding a complex menu or layout.

One design pattern that could be introduced almost anywhere is the singleton. This is because it can be used almost anywhere and its purpose is to provide a global instance of an object.

The singleton pattern

The singleton is easily the simplest of patterns, but it is also one of the most controversial. Many developers think it entirely unnecessary and that declaring a class as static performs the same function with less fuss. Although it is true that the singleton is widely overused when a static class would be the cleaner choice, there are certainly times when one is preferable to the other:

- Use a static class when you want a function performed on a variable you pass to it, for example, calculating the discount value on a price variable
- Use a singleton pattern when you want a complete object, but only one, and you want that object to be available to any part of the program, for example, an object representing the individual user currently logged into an app

The class diagram for the singleton is, as you would imagine, remarkably simple, as you can see here:

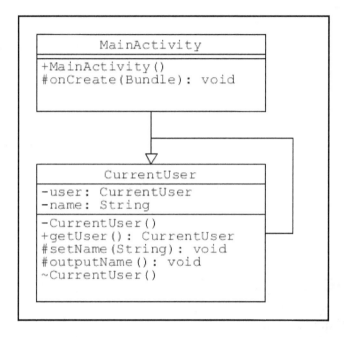

As the preceding diagram suggests, the following example will assume we only have one user logged into our app at any one time and are going to create a singleton object that we can reach from any part of our code.

Android Studio provides for singleton creation under the project explorer's **New** menu, so we can start there. There are only two steps to this demonstration; they are as follows.

1. Add this class to your project:

```
public class CurrentUser {
    private static final String DEBUG_TAG = "tag";
    private String name;

    // Create instance
    private static CurrentUser user = new CurrentUser();

    // Protect class from being instantiated
    private CurrentUser() {
    }

    // Return only instance of user
```

```
    public static CurrentUser getUser() {
        return user;
    }

    // Set name
    protected void setName(String n) {
        name = n;
    }

    // Output user name
    protected void outputName() {
        Log.d(DEBUG_TAG, name);
    }
}
```

2. Test the pattern by adding code like the following to the activity:

```
CurrentUser user = CurrentUser.getUser();
user.setName("Singleton Pattern");
user.outputName();
```

The singleton can be extremely useful, but it is easy to apply it unnecessarily. It is very handy when tasks are asynchronous, such as filing systems, and when we want access to its contents from anywhere in the code, such as the user name in the preceding example.

Summary

Whatever the purpose of an app, users need a familiar way to access it functions. The app-bar and navigation drawer are not only easily understood by the user, but provide great flexibility.

In this chapter, we have seen how to apply two of the most significant input mechanisms available on Android devices and the material patterns that govern their appearance. The SDK, and in particular the design library, make coding these structure both simple and intuitive. Although different from the design patterns we have met so far, material patterns serve a similar function and guide us towards better practice.

The next chapter continues to look into layout design and explores the tools available to us when it comes to putting entire layouts together and how we manage to develop for a wide variety of screen shapes and sizes.

4
Layout Patterns

In the previous chapters, we have looked at the most important patterns used for creating objects and at some of the most used material components. To bring these together, we need to consider the overall layouts an application might need. This allows us to plan our app in greater detail as well as introducing the interesting challenge of designing an app for different-sized screens and orientations. Android makes developing for a variety of screen sizes and shapes very simple and intuitive, and with a minimum of extra coding. We will then conclude by exploring and creating a strategy pattern.

In this chapter, you will learn how to:

- Use relative and linear layouts
- Apply gravity and weight
- Scale weights with weightSum
- Use the percent support library
- Develop layouts for specific screen sizes
- Create a strategy pattern

The Android platform provides a variety of layout classes. These range from the very simple **frame layout** to the quite complex layouts provided by the support library. By far the most widely used and most versatile are the linear and relative layouts.

Linear layouts

Choosing between a relative layout and a linear one is normally very simple. If your components line up from side to side on top of each other, then a **linear layout** is the obvious choice. Although it is quite possible to nest view groups, for more complex layouts, the relative version is often the best choice. This is largely because nesting layouts is resource hungry and deep hierarchies should be avoided where possible. The **relative layout** can be used to create a huge number of intricate layouts with very little need for nesting.

Whichever form best suits our needs, once we begin testing our layouts on screens of different shapes, or even rotate a screen through 90°, we soon see that all the thought that we put into creating components with pleasing proportions is lost. Very often, these issues can be remedied by positioning elements using **gravity** properties and scaling them with the **weight** property.

Weight and gravity

Being able to set position and proportion without having to worry overly about exact screen shape can save us a lot of work. By setting the weight property of components and widgets, we can determine the relative amount of screen width or height an individual component takes up. This is particularly useful when we want most of our widgets set with `wrap_content`, so that they grow as the user needs, but also want one view to take up as much space as is available.

For example, the image in the following layout will shrink appropriately as the text above it grows.

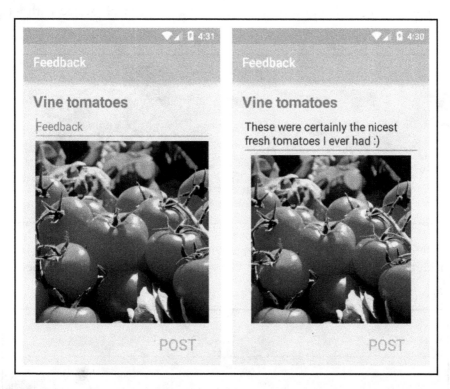

The image view is the only view to have weight applied, the other views all have their height declared with `wrap_content`. As seen here, we have to set the `layout_height` to 0dp here to avoid any internal conflicts when setting the view's height:

```
<ImageView
    android:id="@+id/feedback_image"
    android:layout_width="match_parent"
    android:layout_height="0dp"
    android:layout_weight="1"
    android:contentDescription="@string/content_description"
    android:src="@drawable/tomatoes" />
```

Weight can not only be applied to individual widgets and views, but to view groups and nested layouts.

Automatically filling screen space that is liable to change is very useful, but weight can be applied to more than one view to create layouts where views each consume a specified relative area of an activity. For example, the following images were scaled with weights of 1, 2, 3, and 2.

Although nesting layouts within each other is something to generally avoid, it is often worth considering one or two levels as this can produce some very workable activities. For example:

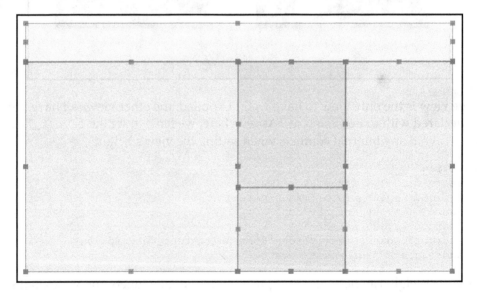

This layout uses only two nested view groups and the use of weight can keep the structure very workable across quite a wide range of form factors. Of course, this layout would look terrible in portrait, but we see how this issue is countered later in the chapter. The XML to generate such a layout would look like this:

```
<FrameLayout
    android:layout_width="match_parent"
    android:layout_height="56dp" />

<LinearLayout
    android:layout_width="match_parent"
    android:layout_height="match_parent"
    android:orientation="horizontal">

    <FrameLayout
        android:layout_width="0dp"
        android:layout_height="match_parent"
        android:layout_weight="2" />

    <LinearLayout
        android:layout_width="0dp"
        android:layout_height="match_parent"
        android:layout_weight="1"
        android:orientation="vertical">

        <FrameLayout
            android:layout_width="match_parent"
            android:layout_height="0dp"
            android:layout_weight="3" />

        <FrameLayout
            android:layout_width="match_parent"
            android:layout_height="0dp"
            android:layout_weight="2" />

    </LinearLayout>

    <FrameLayout
        android:layout_width="0dp"
        android:layout_height="match_parent"
        android:layout_weight="1" />

</LinearLayout>
```

The example above begs an interesting question. What if we do not want to fill the entire width or height of our layout? What if we want some space left? This is easily managed with the **weightSum** property.

To see how weightSum works, add the following highlighted property to the inner linear layout definition in the previous example:

```
<LinearLayout
    android:layout_width="0dp"
    android:layout_height="match_parent"
    android:layout_weight="1"
    android:orientation="vertical"
    android:weightSum="10">
```

By setting a maximum weight for the layout, the inner weights will be set in proportion to this. In this example, a weightSum of 10 sets the inner weights, which are 3 and 2, to 3/10 and 2/10 of the layout height, like so:

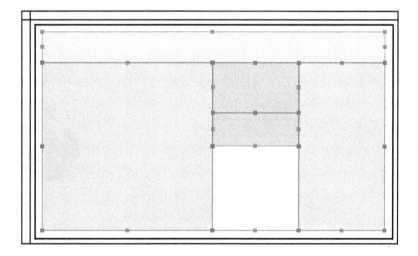

 Note that both the weight and weightSum are floating point properties, and a greater level of accuracy can be achieved with lines such as this: android:weightSum="20.5".

The use of weight is an extremely useful way to make the most of unknown screen sizes and shapes. Another technique for managing overall screen space is to use gravity to position components and their contents.

The **gravity** property is used to justify views and/or their contents. In the example given previously, the following markup was used to position the action at the bottom of the activity:

```
<TextView
    android:id="@+id/action_post"
    android:layout_width="100dp"
    android:layout_height="wrap_content"
    android:layout_gravity="right"
    android:clickable="true"
    android:padding="16dp"
    android:text="@string/action_post"
    android:textColor="@color/colorAccent"
    android:textSize="24sp" />
```

This example demonstrates how `layout_gravity` is used to justify a view (or view group) within its container. The contents of a single view can also be positioned within that view with the `gravity` property, which can be set with something like this:

```
android:layout_gravity="top|left"
```

Ordering our layouts into rows and columns is perhaps the simplest way to consider screen layouts, but it is not the only one. The **relative layout** provides an alternative technique that is based on position rather than proportion. The relative layout also allows us to proportion its content by using the **percent support library**.

Relative layouts

Probably the biggest advantage of the relative layout is the way it can be used to reduce the number of nested view groups when building complex layouts. This works by defining views' positions in accordance to how they are positioned and aligned to each other with properties such as `layout_below` and `layout_toEndOf`. To see how this is done, consider the linear layout of the previous example. We can recreate this as a relative layout with no nested viewgroups, like so:

```
<?xml version="1.0" encoding="utf-8"?>
<RelativeLayout xmlns:android="http://schemas.android.com/apk/res/android"
    android:layout_width="match_parent"
    android:layout_height="match_parent">
```

```xml
    <FrameLayout
        android:id="@+id/header"
        android:layout_width="match_parent"
        android:layout_height="56dp"
        android:layout_alignParentTop="true"
        android:layout_centerHorizontal="true" />

    <FrameLayout
        android:id="@+id/main_panel"
        android:layout_width="320dp"
        android:layout_height="match_parent"
        android:layout_alignParentStart="true"
        android:layout_below="@+id/header" />

    <FrameLayout
        android:id="@+id/center_column_top"
        android:layout_width="160dp"
        android:layout_height="192dp"
        android:layout_below="@+id/header"
        android:layout_toEndOf="@+id/main_panel" />

    <FrameLayout
        android:id="@+id/center_column_bottom"
        android:layout_width="160dp"
        android:layout_height="match_parent"
        android:layout_below="@+id/center_column_top"
        android:layout_toEndOf="@+id/main_panel" />

    <FrameLayout
        android:id="@+id/right_column"
        android:layout_width="match_parent"
        android:layout_height="match_parent"
        android:layout_below="@+id/header"
        android:layout_toEndOf="@+id/center_column_top" />

</RelativeLayout>
```

Despite the obvious advantage that this approach requires no nested view groups, we have to set the individual view's dimensions explicitly, and as soon as we preview the output on different screens, these proportions are soon lost or, at the very least, distorted.

One solution to this issue could be to create separate `dimens.xml` files for different screen configurations, but if we want something that fills a precise percentage of the screen, then we will never be able to guarantee this across every possible device. Fortunately, Android provides a very useful support library.

The percent support library

Being able to define exact proportions for a given component could be something of a problem in a relative layout, as we can only really describe where things are rather than their prominence within the group. Fortunately, the percent library provides **PercentRelativeLayout** to counter this problem.

As with other support libraries, the percent library must be included in the `build.gradle` file:

```
compile 'com.android.support:percent:23.4.0'
```

To create the same layout as previously, we would use the following code:

```
<android.support.percent.PercentRelativeLayout
xmlns:android="http://schemas.android.com/apk/res/android"
    xmlns:app="http://schemas.android.com/apk/res-auto"
    android:layout_width="match_parent"
    android:layout_height="match_parent">

    <FrameLayout
        android:id="@+id/header"
        android:layout_width="match_parent"
        android:layout_height="0dp"
        android:layout_alignParentTop="true"
        android:layout_centerHorizontal="true"
        app:layout_heightPercent="20%" />

    <FrameLayout
        android:id="@+id/main_panel"
        android:layout_width="0dp"
        android:layout_height="match_parent"
        android:layout_alignParentStart="true"
        android:layout_below="@+id/header"
        app:layout_widthPercent="50%" />

    <FrameLayout
        android:id="@+id/center_column_top"
        android:layout_width="0dp"
        android:layout_height="0dp"
```

```
            android:layout_below="@+id/header"
            android:layout_toEndOf="@+id/main_panel"
            app:layout_heightPercent="48%"
            app:layout_widthPercent="25%" />

    <FrameLayout
            android:id="@+id/center_column_bottom"
            android:layout_width="0dp"
            android:layout_height="0dp"
            android:layout_below="@+id/center_column_top"
            android:layout_toEndOf="@+id/main_panel"
            app:layout_heightPercent="32%"
            app:layout_widthPercent="25%" />

    <FrameLayout
            android:id="@+id/right_column"
            android:layout_width="0dp"
            android:layout_height="match_parent"
            android:layout_below="@+id/header"
            android:layout_toEndOf="@+id/center_column_top"
            app:layout_widthPercent="25%" />

</android.support.percent.PercentRelativeLayout>
```

The percent library provides an intuitive and simple way to create proportions that are not easily distorted by being displayed on an untested form factor. These models work very well when tested on other devices with the same orientation. However, once we rotate these layouts through 90°, we can see the problem. Fortunately, the Android SDK allows us to reuse our layout patterns to create alternative versions with the minimum of re-coding. As we might expect, this is achieved by creating designated layout configurations.

Screen rotation

Most, if not all, mobile devices allow for screen reorientation. Many apps, such as video players, are better suited to one orientation than another. Generally speaking, we want our apps to look their best, however rotated.

Most layouts look terrible when translated from portrait to landscape or vice versa. Clearly, we need to create alternatives for these situations. Fortunately, we do not have to start from scratch. The best way to see how this is done is to start with a standard portrait layout like the one here:

This can be recreated with the following code:

```
<android.support.percent.PercentRelativeLayout
xmlns:android="http://schemas.android.com/apk/res/android"
    xmlns:app="http://schemas.android.com/apk/res-auto"
    android:layout_width="match_parent"
    android:layout_height="match_parent">

    <FrameLayout
        android:id="@+id/header"
        android:layout_width="match_parent"
        android:layout_height="0dp"
        android:layout_alignParentTop="true"
        android:layout_centerHorizontal="true"
        android:background="@color/colorPrimary"
        android:elevation="6dp"
        app:layout_heightPercent="10%" />

    <ImageView
        android:id="@+id/main_panel"
        android:layout_width="match_parent"
        android:layout_height="0dp"
        android:layout_alignParentStart="true"
        android:layout_below="@+id/header"
        android:background="@color/colorAccent"
        android:contentDescription="@string/image_description"
        android:elevation="4dp"
        android:scaleType="centerCrop"
        android:src="@drawable/cheese"
        app:layout_heightPercent="40%" />

    <FrameLayout
        android:id="@+id/panel_b"
        android:layout_width="0dp"
        android:layout_height="0dp"
        android:layout_alignParentEnd="true"
        android:layout_below="@+id/main_panel"
        android:background="@color/material_grey_300"
        app:layout_heightPercent="30%"
        app:layout_widthPercent="50%" />

    <FrameLayout
        android:id="@+id/panel_c"
        android:layout_width="0dp"
        android:layout_height="0dp"
        android:layout_alignParentEnd="true"
        android:layout_below="@+id/panel_b"
```

```
        android:background="@color/material_grey_100"
        app:layout_heightPercent="20%"
        app:layout_widthPercent="50%" />

    <FrameLayout
        android:id="@+id/panel_a"
        android:layout_width="0dp"
        android:layout_height="match_parent"
        android:layout_alignParentStart="true"
        android:layout_below="@+id/main_panel"
        android:elevation="4dp"
        app:layout_widthPercent="50%" />

</android.support.percent.PercentRelativeLayout>
```

Again, once this is rotated, it looks very poorly designed. To create an acceptable landscape version, view your layout in design mode and click on the configuration icon in the top-left corner of the design panel and select **Create Landscape Variation**:

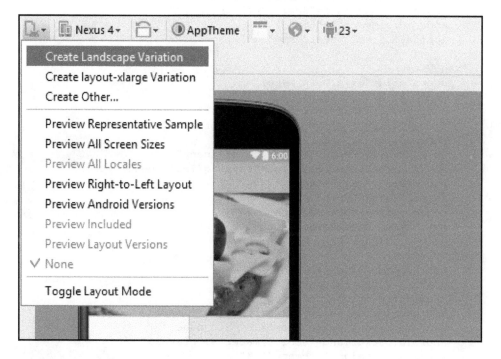

This produces a copy of our file in a folder that will be referred to for its layout definitions whenever the app finds itself in landscape mode. This directory lies alongside the `res/layout` folder and is called `res/layout-land`. It is simply a matter now of rearranging our views to suit this new format, and we can, in fact, use the layout from earlier in the chapter, like so:

```
<android.support.percent.PercentRelativeLayout
xmlns:android="http://schemas.android.com/apk/res/android"
    xmlns:app="http://schemas.android.com/apk/res-auto"
    android:layout_width="match_parent"
    android:layout_height="match_parent">

    <FrameLayout
        android:id="@+id/header"
        android:layout_width="match_parent"
        android:layout_height="0dp"
        android:layout_alignParentTop="true"
        android:layout_centerHorizontal="true"
        android:background="@color/colorPrimary"
        android:elevation="6dp"
        app:layout_heightPercent="15%" />

    <ImageView
        android:id="@+id/main_panel"
        android:layout_width="0dp"
        android:layout_height="match_parent"
        android:layout_alignParentStart="true"
        android:layout_below="@+id/header"
        android:background="@color/colorAccent"
        android:contentDescription="@string/image_description"
        android:elevation="4dp"
        android:scaleType="centerCrop"
        android:src="@drawable/cheese"
        app:layout_widthPercent="50%" />

    <FrameLayout
        android:id="@+id/panel_a"
        android:layout_width="0dp"
        android:layout_height="0dp"
        android:layout_below="@+id/header"
        android:layout_toRightOf="@id/main_panel"
        android:background="@color/material_grey_300"
        app:layout_heightPercent="50%"
        app:layout_widthPercent="25%" />

    <FrameLayout
        android:id="@+id/panel_b"
```

```
        android:layout_width="0dp"
        android:layout_height="0dp"
        android:layout_below="@+id/panel_a"
        android:layout_toRightOf="@id/main_panel"
        android:background="@color/material_grey_100"
        app:layout_heightPercent="35%"
        app:layout_widthPercent="25%" />

    <FrameLayout
        android:id="@+id/panel_c"
        android:layout_width="0dp"
        android:layout_height="match_parent"
        android:layout_alignParentEnd="true"
        android:layout_below="@+id/header"
        android:elevation="4dp"
        app:layout_widthPercent="25%" />

</android.support.percent.PercentRelativeLayout>
```

It only takes a few seconds to apply these changes and create a landscape layout, but there is more we can do here. In particular, we can create layouts designed specifically for larger screens and tablets.

Large screen layouts

When we were creating a landscape version of our layout from the configuration menu, you will no doubt have noticed the **Create layout-xlargeVersion** option, and, as you would imagine, this is used to create layouts suitable for the larger screens of tablets and even TVs.

If you select this option, you will immediately see that our judicious use of the percent library has produced an identical layout, and it is tempting to feel that this layout is unnecessary, but this would be to miss the point. Devices such as a 10" tablet provide a lot more space and rather than just enlarge our layout, we should use this opportunity to provide more content.

In this example, we will just add an extra frame for the xlarge version. This is easily done by adding the following XML and adjusting the height percentage value of some of the other views:

```
<FrameLayout
    android:id="@+id/panel_d"
    android:layout_width="0dp"
    android:layout_height="0dp"
    android:layout_alignParentEnd="true"
    android:layout_below="@+id/panel_c"
    android:background="@color/colorAccent"
    android:elevation="4dp"
    app:layout_heightPercent="30%"
    app:layout_widthPercent="50%" />
```

As well as making the most of larger screens, we can also achieve the opposite for small screens with the small qualifier. This is useful to optimize layouts for small screens by making elements smaller or even removing less important content.

Qualifiers like those we have seen here are very useful but they are still quite broad. Depending on device resolution, we could very easily find the same layout being applied to a large phone and a small tablet. Fortunately, the framework provides a way for us to be more precise when defining our layouts.

Width qualifiers

As developers, we spend a lot of time and energy sourcing and creating great imagery and other media. It is important that we do this work justice and ensure it is displayed at its best. Imagine that you have a layout that deserves to be at least 720 pixels across, to be best appreciated. In such a case, there are two things we can do.

Firstly, we can ensure that our app is only available on devices that have, at least, our desired screen resolution, and this can be done by editing the `AndroidManifest` file, by adding the following tag within the `manifest` element:

```
<supports-screens android:requiresSmallestWidthDp="720" />
```

Usually it would be a shame to make our app unavailable to users of small screens and the times we might do this are rare. Apps designed for large TV screens or for precise photo editing might be the exception. More often we would rather create layouts to suit as many screen sizes as possible and that leads us to our second option.

The Android platform allows us to design layouts for specific screen sizes according to criteria such as **minimum and available width** in pixels. By *minimum*, we mean the narrowest of the two screen dimensions, regardless of orientation. For most devices, this would mean width when viewed in portrait mode and height in landscape mode. The use of *available* width provides another level of flexibility, in that the width is measured according to how a screen is orientated, allowing us to design some very specific layouts. Optimizing layouts according to the smallest width is very simple and is done with qualifiers as before. So a file named:

```
res/layout-sw720dp/activity_main.xml
```

will replace

```
res/layout/activity_main.xml
```

on devices with a shortest side of 720 dp or greater.

Of course, we can create a folder for any size we choose, for example `res/layout-sw600dp`.

This technique is great for designing layouts for large screens regardless of orientation. However, it could be very useful to design a layout that would be applied according to the apparent width based on how a device is oriented at any given moment. This is achieved in a similar manner, by designating directories. To design for available width, use:

```
res/layout-w720dp
```

And to optimize for available height, use:

```
res/layout-h720dp
```

These qualifiers provide very useful techniques for ensuring our designs make the most of available hardware, but there is a slight drawback if we want to develop for devices running Android 3.1 or lower. On these devices, the minimum and available width qualifiers are not available and we have to use `large` and `xlarge` qualifiers. This can lead to two identical layouts, wasting space and adding to our maintenance costs. Thankfully there is a way around this, in the form of layout aliases.

Layout aliases

To demonstrate how layout aliases work we will imagine a simple case where we have just two layouts, our default `activity_main.xml` file which will have just two views and a second layout that we will call `activity_main_large.xml` which will have three views to take advantage of larger screens. To see how this is done, follow these steps:

1. Open the `activity_main` file and provide it with these two views:

```
<ImageView
    android:id="@+id/image_view"
    android:layout_width="match_parent"
    android:layout_height="256dp"
    android:layout_alignParentLeft="true"
    android:layout_alignParentStart="true"
    android:layout_alignParentTop="true"
    android:contentDescription="@string/content_description"
    android:scaleType="fitStart"
    android:src="@drawable/sandwich" />

<TextView
    android:id="@+id/text_view"
    android:layout_width="wrap_content"
    android:layout_height="wrap_content"
    android:layout_below="@+id/image_view"
    android:layout_centerHorizontal="true"
    android:layout_centerVertical="true"
    android:text="@string/text_value"
    android:textAppearance="?android:attr/textAppearanceLarge" />
```

2. Copy this file, call it `activity_main_large` and add the following view to it:

```
<TextView
    android:id="@+id/text_view2"
    android:layout_width="wrap_content"
    android:layout_height="wrap_content"
    android:layout_alignParentEnd="true"
```

```
android:layout_alignParentRight="true"
android:layout_below="@+id/text_view"
android:layout_marginTop="16dp"
android:text="@string/extra_text"
android:textAppearance="?android:attr/textAppearanceMedium" />
```

3. Create two **New | Android resource directories** called `res/values-large` and `res/values-sw720dp`.

4. In the `values-large` folder, create a file called `layout.xml` and complete it like so:

```
<resources>
    <item name="main" type="layout">@layout/activity_main_large</item>
</resources>
```

5. Finally, create an identical file in the `values-sw720dp` folder:

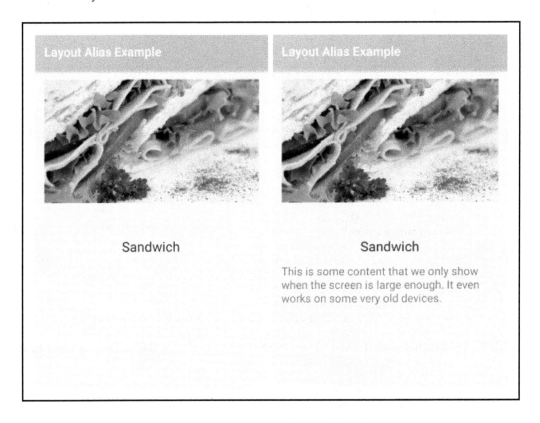

Using layout aliases in this way means we only have to create one large layout and it will be applied to larger screens regardless of which Android platform a device is running.

In this example, we chose `720dp` as our threshold. In most cases this would target 10" tablets and larger. If we wanted our large layout to run on most 7" tablets and large phones, we would use `600dp`, and we can of course select any value that suits our purpose.

There are some, very rare, occasions where we might want to restrict an app to only landscape or portrait. This can be achieved by adding `android:screenOrientation="portrait"` or `android:screenOrientation="landscape"` to the activity tag in the manifest file.

Generally speaking, we should create landscape and portrait layouts for phones, 7" tablets and 10" tablets.

Designing appealing and intuitive layouts is among the most important tasks we face as developers, and the shortcuts introduced here greatly reduce the amount of work we have to do, freeing us up to concentrate on designing attractive applications.

As with the last chapter, we have concentrated on the more practical issue of layout structure, and this is of course prerequisite to further development. However, there are a lot of patterns for us to familiarize ourselves with and the sooner we become familiar with them the better and the more likely that we will identify structures that would benefit from having a pattern applied. Once such pattern that could be applied in situations like those explored in this chapter is the strategy design pattern.

The strategy pattern

The strategy pattern is another widely used and incredibly useful. Its beauty lies in its versatility as it can be applied in numerous situations. Its purpose is to provide a selection of solutions (strategies) to a given problem at runtime. A good example would be an app with a strategy to run different code depending whether the app was being installed on Windows, Mac OS, or Linux. Were the system of designation we used above to design UIs for different devices so efficient, we could easily use a strategy pattern to carry out this task. It would look something like this:

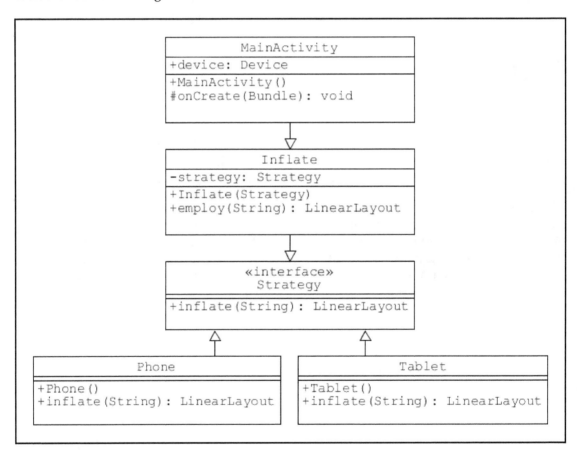

For now, we will step ahead a little and imagine a situation where the users of our sandwich maker app are ready to pay. We will assume three methods: credit card, cash, and a coupon. Those paying cash will simply pay the set price. A little unfairly, those paying by card will be charged a small fee and those with a coupon will get 10% off. We will also use a singleton to represent the basic price before these strategies are applied. Follow these steps to set up a strategy pattern:

1. We start, as is often the case, with an interface:

```
public interface Strategy {

    String processPayment(float price);
}
```

2. Next, create concrete implementations of this interface, like so:

```
public class Cash implements Strategy{

    @Override
    public String processPayment(float price) {

        return String.format("%.2f", price);
    }
}


public class Card implements Strategy{
    ...
        return String.format("%.2f", price + 0.25f);
    ...
}


public class Coupon implements Strategy{
    ...
        return String.format("%.2f", price * 0.9f);
    ...
}
```

3. Now add the following class:

```
public class Payment {
    // Provide context for strategies

    private Strategy strategy;

    public Payment(Strategy strategy) {
        this.strategy = strategy;
    }

    public String employStrategy(float f) {
        return strategy.processPayment(f);
    }
}
```

4. Finally, add the singleton class that is going to provide our basic price:

```
public class BasicPrice {
    private static BasicPrice basicPrice = new BasicPrice();
    private float price;

    // Prevent more than one copy
    private BasicPrice() {
    }

    // Return only instance
    public static BasicPrice getInstance() {
        return basicPrice;
    }

    protected float getPrice() {
        return price;
    }

    protected void setPrice(float v) {
        price = v;
    }
}
```

This is all that we need to do to create the pattern. A singleton was used because the price of the current sandwich is something that needs to only have a single instance and be reached from anywhere in the code. Before we build a UI and test our pattern, let's take a quick look at the strategy class diagram:

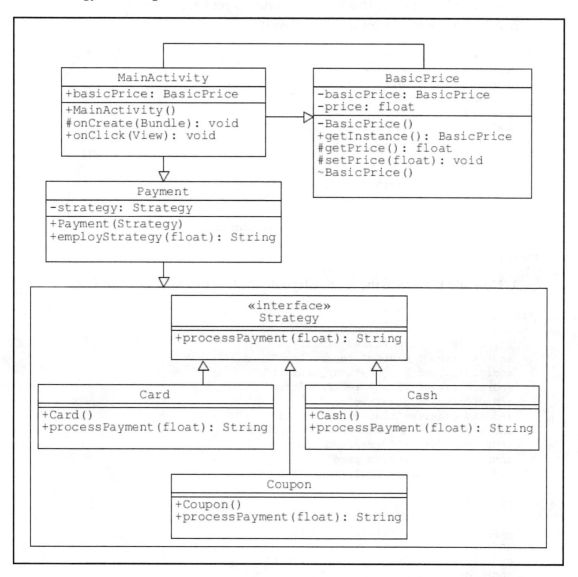

We can see from the diagram that the activity contains an `onClick()` callback. Before we can see how this works, we need to create a layout with three action buttons to test each of our three payment options. Follow these steps to achieve this:

1. Create a layout file with a horizontal linear layout at its root.

2. Add the following view and inner layout:

```
<ImageView
    android:id="@+id/image_view"
    android:layout_width="match_parent"
    android:layout_height="0dp"
    android:layout_weight="1"
    android:scaleType="centerCrop"
    android:src="@drawable/logo" />

<RelativeLayout
    android:layout_width="match_parent"
    android:layout_height="wrap_content"
    android:orientation="horizontal"
    android:paddingTop="@dimen/layout_paddingTop">

</RelativeLayout>
```

3. Now add buttons to the relative layout. The first two look like this:

```
<Button
    android:id="@+id/action_card"
    style="?attr/borderlessButtonStyle"
    android:layout_width="wrap_content"
    android:layout_height="wrap_content"
    android:layout_alignParentEnd="true"
    android:layout_gravity="end"
    android:gravity="center_horizontal"
    android:minWidth="@dimen/action_minWidth"
    android:padding="@dimen/padding"
    android:text="@string/card"
    android:textColor="@color/colorAccent" />

<Button
    android:id="@+id/action_cash"
    style="?attr/borderlessButtonStyle"
    android:layout_width="wrap_content"
    android:layout_height="wrap_content"
    android:layout_gravity="end"
    android:layout_toStartOf="@id/action_card"
    android:gravity="center_horizontal"
    android:minWidth="@dimen/action_minWidth"
```

```
    android:padding="@dimen/padding"
    android:text="@string/cash"
    android:textColor="@color/colorAccent" />
```

4. The third is the same as the second, with the following exceptions:

```
<Button
    android:id="@+id/action_coupon"
    ...
    android:layout_toStartOf="@id/action_cash"
    ...
    android:text="@string/voucher"
    ... />
```

5. Now open the Java activity file extend it so that it implements this listener:

```
public class MainActivity extends AppCompatActivity implements
View.OnClickListener
```

6. Next add the following field:

```
public BasicPrice basicPrice = BasicPrice.getInstance();
```

7. Include these lines on the onCreate() method:

```
// Instantiate action views
Button actionCash = (TextView) findViewById(R.id.action_cash);
Button actionCard = (TextView) findViewById(R.id.action_card);
Button actionCoupon = (TextView) findViewById(R.id.action_coupon);

// Connect to local click listener
actionCash.setOnClickListener(this);
actionCard.setOnClickListener(this);
actionCoupon.setOnClickListener(this);

// Simulate price calculation
basicPrice.setPrice(1.5f);
```

8. Finally add the `onClick()` method, like so:

```
@Override
public void onClick(View view) {
    Payment payment;

    switch (view.getId()) {

        case R.id.action_card:
            payment = new Payment(new Card());
            break;

        case R.id.action_coupon:
            payment = new Payment(new Coupon());
            break;

        default:
            payment = new Payment((new Cash()));
            break;
    }

    // Output price
    String price = new StringBuilder()
            .append("Total cost : $")
            .append(payment.employStrategy(basicPrice.getPrice()))
            .append("c")
            .toString();
    Toast toast = Toast.makeText(this, price, Toast.LENGTH_LONG);
    toast.show();
}
```

We can now test our output on a device or emulator:

The strategy pattern can be applied to many situations and as you develop almost any software, you will come across situations where it can be applied time and again. We will certainly return to it here. Hopefully, introducing it now will help you spot situations where it can be utilized.

Summary

In this chapter, we have seen how to get the most out of Android layouts. This has involved deciding which layout type to use for which purpose, and although there are many others, the linear and relative layouts offer the functionality and flexibility for very many possible layouts. Once a layout has been selected, we can then organize the space with weight and gravity properties. The process of designing layouts for a variety of possible screen sizes was greatly helped by employing the percent library and PercentRelativeLayout.

The biggest challenge a developer faces when designing Android layouts for an enormous number of real world devices our apps might find themselves running on. Fortunately, the use of resource designation makes light work of this.

With a working layout in place, we can move on to look at how we can now use this space to display some useful information. This will take us on to look at how lists and their data are managed by the recycler view, which we shall do in the next chapter.

5
Structural Patterns

So far in this book, we have looked at patterns for holding and returning data and for combining objects into larger ones, but we have not yet considered how we offer a selection of choices to the user.

In planning our sandwich builder app, we would ideally like to offer the customer a wide range of possible ingredients. Probably the best way to present such choices is through a list or, for large collections of data, a series of lists. Android manages these processes very nicely with the **RecyclerView**, which is a list container and manager that replaced the previous ListView. This is not to say that the plain, old list view should never be used, and in cases where all we want is a short, simple text list of a few items, using a recycler view could be considered overkill, and list views are often the way to go. Saying that, the recycler view is far superior particularly when it comes to managing data, keeping the memory footprint small and scrolling smooth, and when contained within a CoordinatorLayout, allows users to drag and drop or swipe and dismiss list items.

To see how all this is done, we will construct an interface, which will consist of a list of ingredients for the user to select from. This will require the RecyclerView for holding the list, which will in turn introduce us to the adapter pattern.

In this chapter, you will learn how to:

- Apply a RecyclerView
- Apply a CoordinatorLayout
- Generate lists
- Translate string resources
- Apply a ViewHolder
- Use a RecyclerView adapter
- Create an adapter design pattern
- Construct bridge design pattern
- Apply facade patterns
- Use patterns to filter data

Generating lists

The RecyclerView is a relatively recent addition and replaces the ListView on older versions. It performs the same functions but manages data far more efficiently, particularly very long lists. The RecyclerView is part of the v7 support library and needs to be compiled in the `build.gradle` file, along with the others shown here:

```
compile 'com.android.support:appcompat-v7:24.1.1'
compile 'com.android.support:design:24.1.1'
compile 'com.android.support:cardview-v7:24.1.1'
compile 'com.android.support:recyclerview-v7:24.1.1'
```

The coordinator layout will form the root layout of the main activity and will look like this:

```
<android.support.design.widget.CoordinatorLayout
xmlns:android="http://schemas.android.com/apk/res/android"
xmlns:app="http://schemas.android.com/apk/res-auto"
android:id="@+id/content"
android:layout_width="match_parent"
android:layout_height="match_parent">
</android.support.design.widget.CoordinatorLayout>
```

The Recycler view can then be placed inside the layout:

```
<android.support.v7.widget.RecyclerView
    android:id="@+id/main_recycler_view"
    android:layout_width="match_parent"
    android:layout_height="match_parent"
    />
```

The RecyclerView provides a dummy list for us, but we will create our list from card views.

List item layouts

It is very tempting to use a card view to display individual items in a list, and one can find many examples of it being used like this. However This practice is not recommended by Google and for good reason. Cards are designed to display content of non-uniform size and the rounded edges and shadows only serve to clutter up the screen. When list items are all the same size and conform to the same layout, then they should appear as simple rectangular layouts, sometimes with a simple divider separating them.

We will be creating complex, interactive list items later in the book, so for now we will just have an image and a string as our item view.

Create a layout file with a horizontal linear layout as its root and place these two views inside it:

```
<ImageView
    android:id="@+id/item_image"
    android:layout_width="@dimen/item_image_size"
    android:layout_height="@dimen/item_image_size"
    android:layout_gravity="center_vertical|end"
    android:layout_margin="@dimen/item_image_margin"
    android:scaleType="fitXY"
    android:src="@drawable/placeholder" />

<TextView
    android:id="@+id/item_name"
    android:layout_width="0dp"
    android:layout_height="wrap_content"
    android:layout_gravity="center_vertical"
    android:layout_weight="1"
    android:paddingBottom="24dp"
    android:paddingStart="@dimen/item_name_paddingStart"
    tools:text="placeholder"
    android:textSize="@dimen/item_name_textSize" />
```

We have used the `tools` namespace here, that should be removed later, just so we can see what our layout looks like without having to compile the entire project:

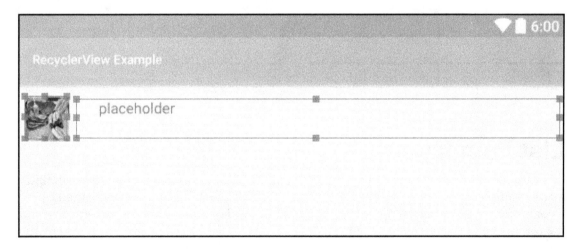

You may have noticed that some margins and padding look different on CardViews when tested on older devices. Rather than resort to creating alternative layout resources, the `card_view:cardUseCompatPadding="true"` property will often resolve this.

The text sizes and margins we applied here are not arbitrary, but those specified by material design guidelines.

Material font sizes

Text size is very important when it comes to material design and only certain sizes are permitted in certain contexts. In the current example, we selected 24sp for the name and 16 for the description. Generally speaking, nearly all the text we will ever display in a material design application will be 12, 14, 16, 20, 24, or 34sp. There is a certain level of flexibility when it comes to selecting which size to use and when, but the following list should provide a good guide:

Display 1: Regular 34sp
Headline: Regular 24sp
Title: Medium 20sp
Subhead: Regular 16sp
Body 2: Medium 14sp
Body 1: Regular 14sp
Caption: Regular 12sp
Button: MEDIUM ALL CAPS 14sp

Connecting data

Android comes equipped with the **SQLite** library, which is a powerful tool for creating and managing complex databases. One could easily fill an entire chapter or even a whole book on the subject. Here we are not dealing with a large collection and it will be simpler and hopefully clearer to create our own data class.

 If you would like to learn more about SQLite, comprehensive documentation can be found at:
`http://developer.android.com/reference/android/database/sqlite/SQLiteDatabase.html`

Later we will create complex data structures, but for now we only need to see how the setup works, so we will create just three items. To add these, create a new Java class called `Filling` and complete like so:

```
public class Filling {
    private int image;
    private int name;

    public Filling(int image, int name) {
        this.image = image;
        this.name = name;
    }
}
```

These can be defined in the main activity like so:

```
static final Filling fillings[] = new Filling[3];
fillings[0] = new Filling(R.drawable.cheese, R.string.cheese);
fillings[1] = new Filling(R.drawable.ham, R.string.ham);
fillings[2] = new Filling(R.drawable.tomato, R.string.tomato);
```

As you can see, we have defined our string resources in the `strings.xml` file:

```
<string name="cheese">Cheese</string>
<string name="ham">Ham</string>
<string name="tomato">Tomato</string>
```

This has two great advantages. Firstly, it allows us to keep view and model separate and secondly, if we were ever to translate our app into other languages this now only requires only an alternative `strings` file. In fact, Android Studio makes this process so simple it is worth taking a moment here to see how it is done.

Translating string resources

Android Studio provides a **translation editor** to simplify the process of providing alternative resources. In exactly the same way that we create designated folders for different screen sizes, we create alternative values directories for different languages. The editor manages this for us and we do not really need to know much about it but it is useful to know that, if we wished to translate our app into Italian, say, then the editor would create a folder named `values-it` and place the alternative `strings.xml` file within it:

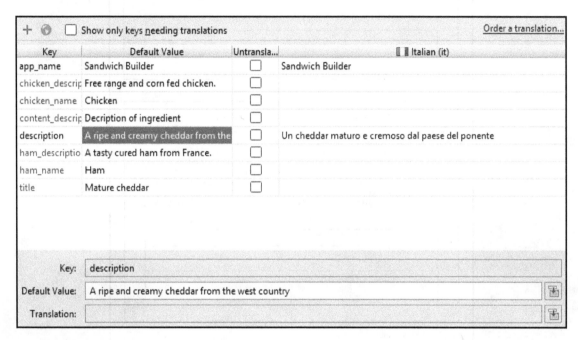

To access the translations editor, simply right-click on the extant `strings.xml` file in the project explorer and select it.

Although the RecyclerView is a fantastic tool for managing and binding data in an efficient manner, it does require quite a bit of setting up. Apart from the view and the data, there are two other elements required to bind the data to our activity, the **LayoutManager** and the **data adapter**.

Adapters and layout managers

RecyclerViews manage their data by using a `RecyclerView.LayoutManager` and a
`RecyclerView.Adapter`. The LayoutManager can be thought of as belonging to the
RecyclerView and it is this that communicates with the adapter, which in turn is bound to
our data in a fashion depicted in the following diagram:

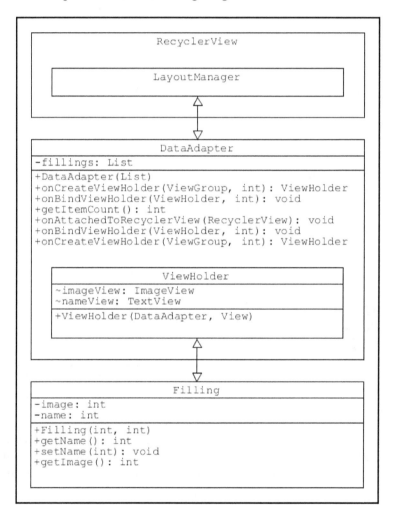

Creating a layout manager is very straightforward. Simply follow these two steps.

1. Open the `MainActivity.Java` file and include the following fields:

```
RecyclerView recyclerView;
DataAdapter adapter;;
```

2. Then add the following lines to the `onCreate()` method:

```
final ArrayList<Filling> fillings = initializeData();
adapter = new DataAdapter(fillings);

recyclerView = (RecyclerView) findViewById(R.id.recycler_view);
recyclerView.setHasFixedSize(true);
recyclerView.setLayoutManager(new LinearLayoutManager(this));
recyclerView.setAdapter(adapter);
```

This code is simple to understand but the purpose of the `RecyclerView.setHasFixedSize(true)` command my need some explanation. If we know in advance that our list is always going to be the same length, then this call will make the management of the list far more efficient.

To create the adapter, follow these steps:

1. Create a new Java class called `DataAdapter` and have it extend `RecyclerView.Adapter<RecyclerViewAdapter.ViewHolder`.
2. This will generate and error, click on the red quick-fix icon and implement the methods suggested.
3. These three methods should be filled out as seen here:

```
// Inflate recycler view
@Override
public DataAdapter.ViewHolder onCreateViewHolder(ViewGroup parent, int
viewType) {
    Context context = parent.getContext();
    LayoutInflater inflater = LayoutInflater.from(context);

    View v = inflater.inflate(R.layout.item, parent, false);
    return new ViewHolder(v);
    }

// Display data
@Override
public void onBindViewHolder(DataAdapter.ViewHolder holder, int position) {
    Filling filling = fillings.get(position);
```

```
    ImageView imageView = holder.imageView;
    imageView.setImageResource(filling.getImage());

    TextView textView = holder.nameView;
    textView.setText(filling.getName());
}

@Override
@Overridepublic int getItemCount() {    return fillings.size();}
```

4. And finally, the ViewHolder:

```
public class ViewHolder extends RecyclerView.ViewHolder {
    ImageView imageView;
    TextView nameView;

    public ViewHolder(View itemView) {
        super(itemView);
        imageView = (ImageView) itemView.findViewById(R.id.item_image);
        nameView = (TextView) itemView.findViewById(R.id.item_name);
    }
}
```

The **ViewHolder** speeds up long lists by only making one call to `findViewById()`, which is a resource hungry process.

The example can now be run on an emulator or handset and will have an output similar to that seen here:

Obviously, we would want far more than three fillings but it is easy to see from this example how we could add as many more as we wished.

The example we have worked through here explains enough of how the RecyclerView works to be able to implement one in a variety of situations. We used a LinearLayoutManager with one here to create our list, but there is also a **GridLayoutManager** and a **StaggeredGridLayoutManager** that work in a very similar fashion.

The adapter pattern

In the example we have been studying here, we used an adapter pattern to connect our data with our layout in the form of our `DataAdapter`. This is a ready-made adapter and although it is clear how it works, it teaches us nothing about the structure of the adapter or how to construct one ourselves.

There are many cases where Android provides built in patterns, which is very useful, but there will often be times when we need an adapter for classes we have created ourselves, and we will now see how this is done and also how to create the associated design pattern, the bridge. It is best to begin by looking at these patterns conceptually.

The purpose of the adapter is perhaps the easiest to understand. A good analogy would be the physical adapters we use when we take electronic devices to other countries where their power outlets work on different voltages and frequencies. The adapter has two faces, one to accept our plug and one to fit the socket. Some adapters are even smart enough to accept more than one configuration, and this is exactly how software adapters work.

There are many occasions when we have interfaces that do not match up in the same way as plugs do not line up with foreign sockets and the adapter is one of the most widely relied on design patterns. We saw earlier that Android APIs themselves make use of them.

One way to solve the problem of incompatible interfaces is to change the interfaces themselves, but this can result in some very messy code and spaghetti-like connections between classes. Adapters solve this problem and also allow for us to make wide scale changes to our software without really disrupting overall structures.

Imagine that our sandwich app has launched and is doing fine, but then the offices we deliver to change their floor plans and go from small offices to an open plan structure. Previously we had used fields for building, floor, office, and desk to locate customers, but now the office field makes no sense and we have to redesign accordingly.

If our application is at all complex, there will no doubt be many references and uses of location classes and rewriting them all could be a time-consuming business. Fortunately, the adapter pattern means we can adapt to this change with very little fuss.

Here is the original location interface:

```java
public interface OldLocation {

    String getBuilding();
    void setBuilding(String building);

    int getFloor();
    void setFloor(int floor);

    String getOffice();
    void setOffice(String office);

    int getDesk();
    void setDesk(int desk);
}
```

Here is how it would be implemented:

```java
public class CustomerLocation implements OldLocation {
    String building;
    int floor;
    String office;
    int desk;

    @Override
    public String getBuilding() { return building; }

    @Override
```

```
    public void setBuilding(String building) {
        this.building = building;
    }

    @Override
    public int getFloor() { return floor; }

    @Override
    public void setFloor(int floor) {
        this.floor = floor;
    }

    @Override
    public String getOffice() { return office; }

    @Override
    public void setOffice(String office) {
        this.office = office;
    }

    @Override
    public int getDesk() { return desk; }

    @Override
    public void setDesk(int desk) {
        this.desk = desk;
    }
}
```

Assuming that these classes already exist and it is these that we wish to adapt, it takes just the adapter class and some test code to convert a whole app from the old system to the new:

1. The adapter class:

```
public class Adapter implements NewLocation {
    final OldLocation oldLocation;

    String building;
    int floor;
    int desk;

    // Wrap in old interface
    public Adapter(OldLocation oldLocation) {
        this.oldLocation = oldLocation;
        setBuilding(this.oldLocation.getBuilding());
        setFloor(this.oldLocation.getFloor());
        setDesk(this.oldLocation.getDesk());
    }
```

```
        @Override
        public String getBuilding() { return building; }

        @Override
        public void setBuilding(String building) {
            this.building = building;
        }

        @Override
        public int getFloor() { return floor; }

        @Override
        public void setFloor(int floor) {
            this.floor = floor;
        }

        @Override
        public int getDesk() { return desk; }

        @Override
        public void setDesk(int desk) {
            this.desk = desk;
        }
    }
```

2. The test code:

```
TextView textView = (TextView)findViewById(R.id.text_view);

OldLocation oldLocation = new CustomerLocation();
oldLocation.setBuilding("Town Hall");
oldLocation.setFloor(3);
oldLocation.setDesk(14);

NewLocation newLocation = new Adapter(oldLocation);

textView.setText(new StringBuilder()
        .append(newLocation.getBuilding())
        .append(", floor ")
        .append(newLocation.getFloor())
        .append(", desk ")
        .append(newLocation.getDesk())
        .toString());
```

Despite its usefulness, the adapter pattern has a very simple structure, as can be seen in the diagram here:

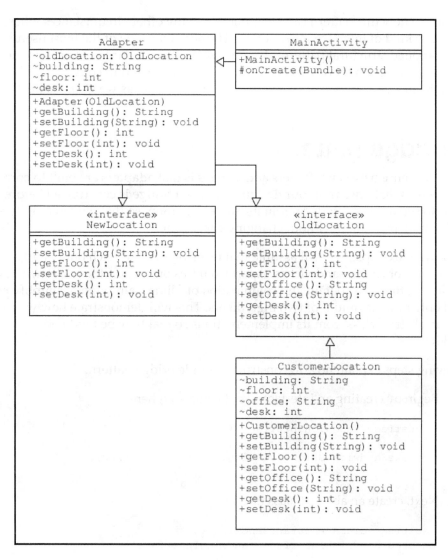

The key to the adapter pattern is the way that the adapter class implements the new interface and wraps the old one.

It is easy to see how this pattern can be applied in many other circumstances where we need to convert one kind of interface into another. The adapter is one of the most useful and frequently applied structural patterns. In some ways, it is similar to the next pattern we will encounter, the bridge, in that they both have a class that is used to convert interfaces. However, the bridge pattern serves a very different function, as we shall see next.

The bridge pattern

The main difference between adapters and bridges is that adapters are built to correct incompatibilities that arise from our design, whereas a bridge is constructed before, and its purpose is to separate an interface from its implementation, so that we can modify and even replace the implementation without changing client code.

In the following example, we will assume that users of our sandwich builder app will have a choice of open or closed sandwiches. Apart from this one factor, these sandwiches are identical in that they can contain any combination of fillings, although to simplify matters, there will only be a maximum of two ingredients. This will demonstrate how we can decouple an abstract class from its implementations so that they be modified independently.

The following steps explain how to construct a simple bridge pattern:

1. Begin by creating an interface like the one seen here:

```
public interface SandwichInterface {

    void makeSandwich(String filling1, String filling2);
}
```

2. Next, create an abstract class like so:

```
public abstract class AbstractSandwich {
    protected SandwichInterface sandwichInterface;

    protected AbstractSandwich(SandwichInterface sandwichInterface) {
        this.sandwichInterface = sandwichInterface;
    }

    public abstract void make();
}
```

3. Now extend this class like this:

```
public class Sandwich extends AbstractSandwich {
    private String filling1, filling2;

    public Sandwich(String filling1, String filling2, SandwichInterface
sandwichInterface) {
        super(sandwichInterface);
        this.filling1 = filling1;
        this.filling2 = filling2;
    }

    @Override
    public void make() {
        sandwichInterface.makeSandwich(filling1, filling2);
    }
}
```

4. And then create two concrete classes to represent our choice of sandwich:

```
public class Open implements SandwichInterface {
    private static final String DEBUG_TAG = "tag";

    @Override
    public void makeSandwich(String filling1, String filling2) {
        Log.d(DEBUG_TAG, "Open sandwich " + filling1 + filling2);
    }
}

public class Closed implements SandwichInterface {
    private static final String DEBUG_TAG = "tag";

    @Override
    public void makeSandwich(String filling1, String filling2) {
        Log.d(DEBUG_TAG, "Closed sandwich " + filling1 + filling2);
    }
}
```

5. This pattern can now be tested by adding these lines to the client code:

```
AbstractSandwich openSandwich = new Sandwich("Cheese ", "Tomato", new
Open());
openSandwich.make();

AbstractSandwich closedSandwich = new Sandwich("Ham ", "Eggs", new
Closed());
closedSandwich.make();
```

6. The output in the debug screen will then match this:

```
D/tag: Open sandwich Cheese Tomato
D/tag: Closed sandwich Ham Eggs
```

What this demonstrates is how the pattern allows us to make our sandwich in different ways using the same abstract class method but different bridge implementer classes.

Both adapters and bridges work by creating clean structures that we can use to unify or separate classes and interfaces to solve structural incompatibilities as they arise, or anticipate them during planning. When viewed diagrammatically, the differences between the two become more apparent:

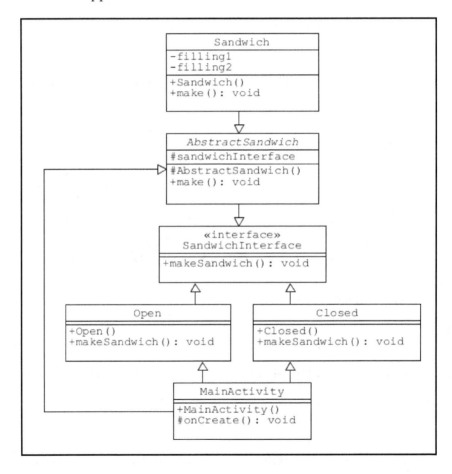

Most structural patterns (and design patterns in general) rely on creating these extra layers to clarify code. Simplifying complex structure is without doubt a design pattern's greatest asset, and very few patterns help us simplify code more than the facade pattern.

The facade pattern

The facade pattern is perhaps one of the simplest of the structural pattern to understand and create. As its name suggests, it act as a face that sits in front of a complex system. When programming client code, we never have to concern ourselves with the complex logic of the rest of our system, if we have a facade to represent it. All we have to do is deal with the facade itself, and this means we can devise facades to maximize simplicity.

Think of the facade pattern like the simple keypad you might find on a typical vending machine. Vending machines are very complex systems, combining all manner of mechanical and physical components. However, to operate one, all we need is to know how to punch in a number or two on its keypad. The keypad is the facade and it hides all the complexity behind it. We can demonstrate this by considering the imaginary vending machine, outlined in the following steps:

1. Start by creating the following interface:

```
public interface Product {

    int dispense();
}
```

2. Next, add three concrete implementations of this, like so:

```
public class Crisps implements Product {

    @Override
    public int dispense() {
        return R.drawable.crisps;
    }
}

public class Drink implements Product {
    ...
        return R.drawable.drink;
    ...
}

public class Fruit implements Product {
    ...
```

```
        return R.drawable.fruit;
    ...
}
```

3. Now add the facade class:

```
public class Facade {
    private Product crisps;
    private Product fruit;
    private Product drink;

    public Facade() {
        crisps = new Crisps();
        fruit = new Fruit();
        drink = new Drink();
    }

    public int dispenseCrisps() {
        return crisps.dispense();
    }

    public int dispenseFruit() {
        return fruit.dispense();
    }

    public int dispenseDrink() {
        return drink.dispense();
    }
}
```

4. Place suitable images in the appropriate drawable directories.
5. Create a simple layout file with an image view similar to this one:

```
<ImageView
    android:id="@+id/image_view"
    android:layout_width="match_parent"
    android:layout_height="match_parent" />
```

6. Add an `ImageView` to the activity class:

```
ImageView imageView = (ImageView) findViewById(R.id.image_view);
```

7. Create a facade:

```
Facade facade = new Facade();
```

8. Then test output with calls like the one here:

```
imageView.setImageResource(facade.dispenseCrisps());
```

This constitutes our facade pattern. It is pretty simple to visualize:

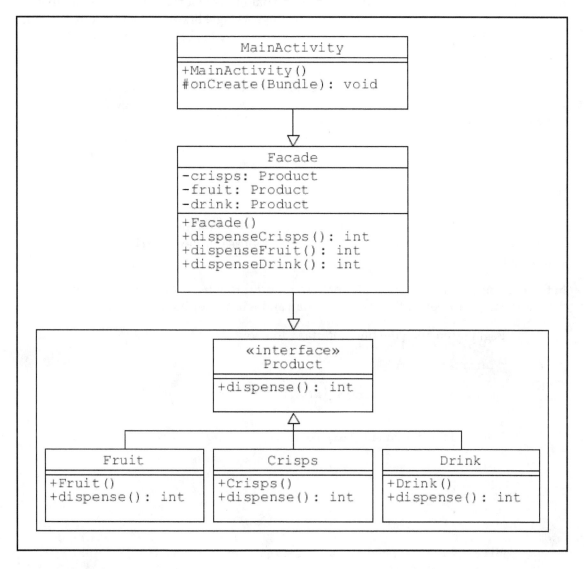

Of course, the facade pattern in this example might seem pointless. The `dispense()` method does nothing more than display an image, and requires no simplification. However, in a more realistic simulation, the dispensing process would involve all manner of calls and checks, change has to be calculated, stock availability checked, and any number of servos need setting into action. The beauty of the facade pattern is that if we were to put all these procedures into place, we would not have to change a single line in our client code or facade class. A single call to `dispenseDrink()` will have the correct result, no matter how complex the logic behind it.

Despite being very simple, the facade pattern is immensely useful in many situations where we want to present a simple and orderly interface for a complicated system. Far less simple but equally useful is the criteria (or filter) pattern, which allows us to interrogate complex data structures.

The criteria pattern

The criteria design pattern provides a clear and concise technique for filtering objects according to set criteria. It can be a very powerful tool as this next exercise will demonstrate.

In this example, we will apply a filter pattern to sort through a list of ingredients and filter them according to whether they are vegetarian and where they are produced:

1. Begin by creating the filter interface, like so:

```
public interface Filter {

    List<Ingredient> meetCriteria(List<Ingredient> ingredients);
}
```

2. Next add the ingredient class, like this:

```
public class Ingredient {

    String name;
    String local;
    boolean vegetarian;

    public Ingredient(String name, String local, boolean vegetarian){
        this.name = name;
        this.local = local;
        this.vegetarian = vegetarian;
    }
```

```
    public String getName() {
        return name;
    }

    public String getLocal() {
        return local;
    }

    public boolean isVegetarian(){
        return vegetarian;
    }
}
```

3. Now implement the filter to meet the vegetarian criteria:

```
public class VegetarianFilter implements Filter {

    @Override
    public List<Ingredient> meetCriteria(List<Ingredient> ingredients) {
        List<Ingredient> vegetarian = new ArrayList<Ingredient>();

        for (Ingredient ingredient : ingredients) {
            if (ingredient.isVegetarian()) {
                vegetarian.add(ingredient);
            }
        }
        return vegetarian;
    }
}
```

4. Then, add a filter to test for local produce:

```
public class LocalFilter implements Filter {

    @Override
    public List<Ingredient> meetCriteria(List<Ingredient> ingredients) {
        List<Ingredient> local = new ArrayList<Ingredient>();

        for (Ingredient ingredient : ingredients) {
            if (Objects.equals(ingredient.getLocal(), "Locally produced"))
{
                local.add(ingredient);
            }
        }
        return local;
    }
}
```

5. And one for non-local ingredients:

```
public class NonLocalFilter implements Filter {

    @Override
    public List<Ingredient> meetCriteria(List<Ingredient> ingredients) {
        List<Ingredient> nonLocal = new ArrayList<Ingredient>();

        for (Ingredient ingredient : ingredients) {
            if (ingredient.getLocal() != "Locally produced") {
                nonLocal.add(ingredient);
            }
        }
        return nonLocal;
    }
}
```

6. Now we need to include an AND criteria filter:

```
public class AndCriteria implements Filter {
    Filter criteria;
    Filter otherCriteria;

    public AndCriteria(Filter criteria, Filter otherCriteria) {
        this.criteria = criteria;
        this.otherCriteria = otherCriteria;
    }

    @Override
    public List<Ingredient> meetCriteria(List<Ingredient> ingredients) {
        List<Ingredient> firstCriteria =
criteria.meetCriteria(ingredients);
        return otherCriteria.meetCriteria(firstCriteria);
    }
}
```

7. Followed by an OR criteria:

```
public class OrCriteria implements Filter {
    Filter criteria;
    Filter otherCriteria;

    public OrCriteria(Filter criteria, Filter otherCriteria) {
        this.criteria = criteria;
        this.otherCriteria = otherCriteria;
    }

    @Override
```

```
    public List<Ingredient> meetCriteria(List<Ingredient> ingredients) {
        List<Ingredient> firstCriteria =
criteria.meetCriteria(ingredients);
        List<Ingredient> nextCriteria =
otherCriteria.meetCriteria(ingredients);

        for (Ingredient ingredient : nextCriteria) {
            if (!firstCriteria.contains(ingredient)) {
                firstCriteria.add(ingredient);
            }
        }
        return firstCriteria;
    }
}
```

8. Now, add a small data set along these lines:

```
List<Ingredient> ingredients = new ArrayList<Ingredient>();

ingredients.add(new Ingredient("Cheddar", "Locally produced", true));
ingredients.add(new Ingredient("Ham", "Cheshire", false));
ingredients.add(new Ingredient("Tomato", "Kent", true));
ingredients.add(new Ingredient("Turkey", "Locally produced", false));
```

9. In the main activity, create the following filters:

```
Filter local = new LocalFilter();
Filter nonLocal = new NonLocalFilter();
Filter vegetarian = new VegetarianFilter();
Filter localAndVegetarian = new AndCriteria(local, vegetarian);
Filter localOrVegetarian = new OrCriteria(local, vegetarian);
```

10. Create a simple layout with a basic text view.

11. Add the following method to the main activity:

```
public void printIngredients(List<Ingredient> ingredients, String header) {

    textView.append(header);

    for (Ingredient ingredient : ingredients) {
        textView.append(new StringBuilder()
                .append(ingredient.getName())
                .append(" ")
                .append(ingredient.getLocal())
                .append("\n")
                .toString());
    }
}
```

12. The pattern can now be tested with calls like those here:

```
printIngredients(local.meetCriteria(ingredients),
"LOCAL:\n");
printIngredients(nonLocal.meetCriteria(ingredients),
"\nNOT LOCAL:\n");
printIngredients(vegetarian.meetCriteria(ingredients),
"\nVEGETARIAN:\n");
printIngredients(localAndVegetarian.meetCriteria(ingredients),
"\nLOCAL VEGETARIAN:\n");
printIngredients(localOrVegetarian.meetCriteria(ingredients),
"\nENVIRONMENTALLY FRIENDLY:\n");
```

Testing the pattern on a device should produce this output:

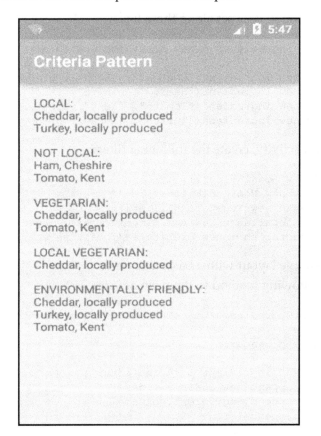

We only applied a few simple criteria here, but we could easily have included information regarding allergies, calories, price, and any others we chose, along with the appropriate filters. It is this ability to create a single criterion from multiple ones that makes this pattern so useful and versatile. It can be viewed visually like this:

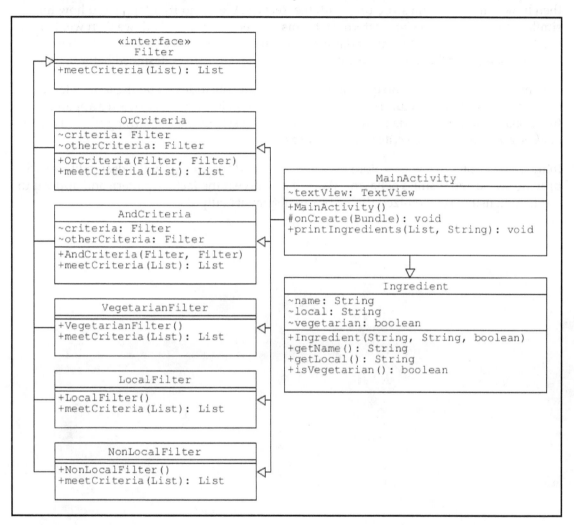

The filter pattern, like many others, does not do anything that we have not done before. Instead it shows another way to perform familiar and common tasks, such as filtering data according to specific criteria. Providing we pick the right pattern for the right task, these tried and tested structures make best practice almost inevitable.

Summary

In this chapter, we have covered some of the most frequently applied and most useful of the structural patterns. We began by seeing how the framework separates model from view and then how to manage data structures with the RecyclerView and its adapter and how this is similar to the adapter design pattern. With this connection established, we then went on to create an example of how we use adapters to counter the inevitable incompatibilities between objects, unlike the bridge pattern we built next, which is designed beforehand.

Having started the chapter on quite a practical note, it concludes by taking a close look at two other important structural patterns, the facade, for simplifying structures apparent functionality and the criteria pattern, which works on sets of data, returning filtered sets of objects, applying multiple criteria as simply as we might one.

In the next chapter, we will explore user interfaces and how to apply the design library to provide swipe and dismiss behavior. We will also revisit the factory pattern and apply it to our layout, using a customized dialog box to display its output.

6
Activating Patterns

The chapters up to this point have served as an extended introduction, exploring the practicalities of Android development and the theory of design pattern application. We have covered many of the fundamental components of an Android app and seen how some of the most useful patterns are made, but we have not yet put the two together.

In this chapter, we will build one of the main sections of our app, an ingredient selection menu. This will involve a scrollable list of fillings that can be selected, expanded, and dismissed. On the way, we will also take a look at the collapsible toolbar and one or two other handy support library features, adding functionality to action buttons, a floating action button, and an alert dialog.

At the heart of this code, we will apply a simple factory pattern to create each ingredient. This will demonstrate nicely how this pattern hides creational logic from client classes. In this chapter, we will create only a single example of a filling type, to see how it is done, but the same structures and processes will be used later as more complexity is added. This will lead us to explore recycler view formats and decoration, such as grid layouts and dividers.

We will then move on to generate and customize an alert dialog from the clicking of a button. This will require an inbuilt builder pattern and lead us on to see how we can create a builder pattern of our own for inflating layouts.

In this chapter, you will learn how to:

- Create an app-bar layout
- Apply a collapsing toolbar
- Control scrolling behavior
- Include nested scroll views
- Apply a data factory
- Create a list item view
- Convert a text view into a button
- Apply grid layouts
- Add divider decoration
- Configure action icons
- Create an alert dialog
- Customize dialogs
- Add a second activity
- Apply swipe and dismiss behavior
- Create a layout builder pattern
- Create a layout at runtime

Users of our app will need some way of selecting ingredients. We could of course present them with one long list, but this would be cumbersome and unattractive. Clearly, we need to split our ingredients into categories. In the following examples, we will concentrate on just one of these groups as this will help simplify the underlying processes for later on, when we will consider more complex scenarios. We will begin by creating the necessary layouts, starting with the collapsing toolbar layout.

Collapsing toolbars

Toolbars that slide out of the way conveniently are a common feature of material design UIs, and provide an elegant and clever way to make good use of the limited space available on phones and even laptops.

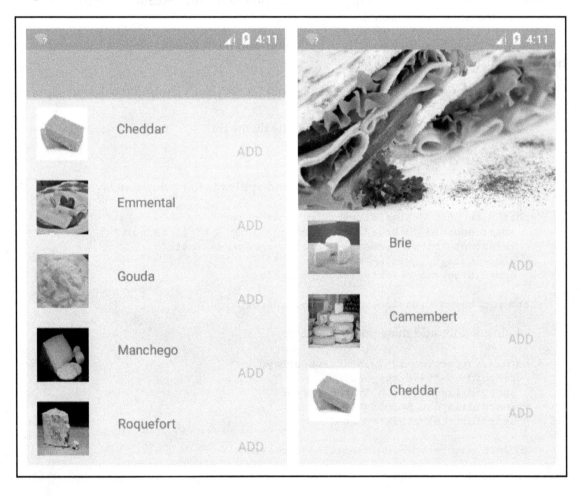

As you would imagine the **CollapsingToolbarLayout** is part of the design support library. It is intended as a child of the **AppBarLayout**, which is a linear layout, designed specifically for material design features.

Collapsing toolbars help manage space elegantly and also provide a good opportunity to display attractive graphics and help promote our product. They take little time to implement and are easily adapted.

The best way to see how they work is to build one, and the following steps demonstrate how to do this:

1. Start a new project and include the both the recycler view and the design support libraries.
2. Remove the action bar by changing the theme to:

```
Theme.AppCompat.Light.NoActionBar
```

3. Open the `activity_main.xml` file and apply the following root layout:

```
<android.support.design.widget.CoordinatorLayout
    xmlns:android="http://schemas.android.com/apk/res/android"
    xmlns:app="http://schemas.android.com/apk/res-auto"
    android:layout_width="match_parent"
    android:layout_height="match_parent">

</android.support.design.widget.CoordinatorLayout>
```

4. Inside this, add this `AppBarLayout`:

```
<android.support.design.widget.AppBarLayout
    android:id="@+id/app_bar"
    android:layout_width="match_parent"
    android:layout_height="wrap_content"
    android:fitsSystemWindows="true">

</android.support.design.widget.AppBarLayout>
```

5. Place this `CollapsingToolbarLayout` inside the app-bar:

```
<android.support.design.widget.CollapsingToolbarLayout
    android:id="@+id/collapsing_toolbar"
    android:layout_width="match_parent"
    android:layout_height="wrap_content"
    android:fitsSystemWindows="true"
    app:contentScrim="?attr/colorPrimary"
app:layout_scrollFlags="scroll|exitUntilCollapsed|enterAlwaysCollapsed">

</android.support.design.widget.CollapsingToolbarLayout>
```

6. The contents of the collapsing toolbar are the following two views:

```
<ImageView
    android:id="@+id/toolbar_image"
    android:layout_width="match_parent"
    android:layout_height="match_parent"
    android:fitsSystemWindows="true"
    android:scaleType="centerCrop"
    android:src="@drawable/some_drawable"
    app:layout_collapseMode="parallax" />

<android.support.v7.widget.Toolbar
    android:id="@+id/toolbar"
    android:layout_width="match_parent"
    android:layout_height="?attr/actionBarSize"
    app:layout_collapseMode="pin" />
```

7. Now, below the app-bar layout, add this recycler view:

```
<android.support.v7.widget.RecyclerView
    android:id="@+id/recycler_view"
    android:layout_width="match_parent"
    android:layout_height="match_parent"
    android:scrollbars="vertical"
    app:layout_behavior="@string/appbar_scrolling_view_behavior" />
```

8. Finally, add this floating action button:

```
<android.support.design.widget.FloatingActionButton
    android:id="@+id/fab"
    android:layout_width="wrap_content"
    android:layout_height="wrap_content"
    android:layout_marginEnd="@dimen/fab_margin_end"
    app:layout_anchor="@id/app_bar"
    app:layout_anchorGravity="bottom|end" />
```

 It is possible, and often desirable, to set the status bar to translucent so that our app-bar image can be seen behind it. This is achieved by adding the following two items to the styles.xml files:

```
<item name="android:windowDrawsSystemBarBackgrounds">true</item>
<item name="android:statusBarColor">@android:color/transparent</item>
```

We have already encountered the coordinator layout in a previous chapter and seen how it facilitates many material design functions. The `AppBarLayout` does a similar thing and is generally used as a container for collapsing toolbars.

The **CollapsingToolbarLayout**, on the other hand, needs one or two things explaining. Firstly, the use of `android:layout_height="wrap_content"` will produce different effects, depending on the height of the image its ImageView contains. This is done so that when we design alternative layouts for different screen sizes and densities, we can simply scale this image accordingly. Here it is configured for a small (480 x 854dp) 240dpi device and is 192dp tall. We could of course have set layout height in dp and scaled this value in the various `dimens.xml` files. We would however, still have had to scale the image, so this method kills two birds with one stone.

The other interesting point about the collapsing toolbar layout is the way we can control how it scrolls, and as you would imagine, this is dealt with by the **layout_scrollFlags** attribute. Here we used `scroll`, `exitUntilCollapsed`, `enterAlwaysCollapsed`. This means that the toolbar never disappears from the top of the screen and that the toolbar does not expand until the list can be scrolled no further down.

There are five scroll flags, and they are:

- `scroll` – Enables scrolling
- `exitUntilCollapsed` – Prevents the toolbar from disappearing when scrolling up (omit to lose the toolbar until scrolling down)
- `enterAlways` – Toolbar expands whenever the list scrolls down
- `enterAlwaysCollapsed` – Toolbar only expands from top of the list
- `snap` – Toolbar snaps into place rather than gliding

The image view within the collapsing toolbar is almost identical to any other image view we might have seen, apart from maybe the `layout_collapseMode` attribute. This has two possible settings, `pin` and `parallax`:

- `pin` – The list and toolbar move together
- `parallax` – The list and toolbar move separately

The best way to appreciate these effects is simply to try them out. We could also have applied either of these layout collapse modes on the toolbar beneath the image, but as we want our toolbar to remain on screen, we need not concern ourselves with its collapsing behavior.

The recycler view that will contain our data here is different in only one respect from the one we used earlier in the book. That is the inclusion of the line:

```
app:layout_behavior="@string/appbar_scrolling_view_behavior"
```

This attribute is all we have to add to any view or view group that sits below the app bar, to allow the two to coordinate their scrolling behavior.

These simple classes save us a great deal of work when it comes to implementing material design and leave us to concentrate on providing functionality. Apart from the size of the image, very little refactoring is required to create a layout that works on a large number of possible devices.

Although we are using a recycler view here, it is quite possible to put any number of views and view groups below the app-bar. Providing that they possess the `app:layout_behavior="@string/appbar_scrolling_view_behavior"` attribute, they will move in concord with the bar. There is one layout that particularly suits this purpose and that is the **NestedScrollView**. By way of example, it looks like this:

```
<android.support.v4.widget.NestedScrollView
    android:layout_width="match_parent"
    android:layout_height="match_parent"
    app:layout_behavior="@string/appbar_scrolling_view_behavior">

    <TextView
        android:id="@+id/nested_text"
        android:layout_width="match_parent"
        android:layout_height="wrap_content"
        android:padding="@dimen/nested_text_padding"
        android:text="@string/some_text"
        android:textSize="@dimen/nested_text_textSize" />

</android.support.v4.widget.NestedScrollView>
```

The next logical step is to create a layout for populating the recycler view, but first we need to prepare the data. In this chapter, we will develop an application component responsible for presenting the user with a list of ingredients of a particular category, in this case cheese. We will use the **factory pattern** to create these objects.

Applying a data factory pattern

In this section, we will apply a factory pattern to create objects of type *cheese*. This will in turn implement a *filling* interface. Each object will consist of several properties such as price and calorific value. Some of these values will be presented in our list items and others will be available only through an expanded view or accessible only via code.

One of the few disadvantages of design patterns is the large number of classes that soon accumulate. For this reason, before beginning the following exercise, create a new package inside the `java` directory, called `fillings`.

Follow these steps to generate our cheese factory:

1. Create a new interface called `Filling` in the `fillings` package and complete it like so:

```
public interface Filling {

    String getName();
    int getImage();
    int getKcal();
    boolean isVeg();
    int getPrice();
}
```

2. Next, create an abstract class that implements `Filling`, called `Cheese`, like this:

```
public abstract class Cheese implements Filling {
    private String name;
    private int image;
    private String description;
    private int kcal;
    private boolean vegetarian;
    private int price;

    public Cheese() {
    }

    public abstract String getName();

    public abstract int getImage();

    public abstract int getKcal();
```

```
public abstract boolean getVeg();

public abstract int getPrice();
}
```

3. Create a concrete class called `Cheddar`, like the one here:

```java
public class Cheddar extends Cheese implements Filling {

    @Override
    public String getName() {
        return "Cheddar";
    }

    @Override
    public int getImage() {
        return R.drawable.cheddar;
    }

    @Override
    public int getKcal() {
        return 130;
    }

    @Override
    public boolean getVeg() {
        return true;
    }

    @Override
    public int getPrice() {
        return 75;
    }
}
```

4. Create several other `Cheese` classes, along the lines of the `Cheddar`.

Having created the factory, we need a way to represent each cheese. For this, we will create an item layout.

Positioning item layouts

To keep the interface clean, we will create a very simple item for our recycler view list. It will contain just an image, a string, and an action button for the user to add the ingredient to their sandwich.

The initial item layout will look like this:

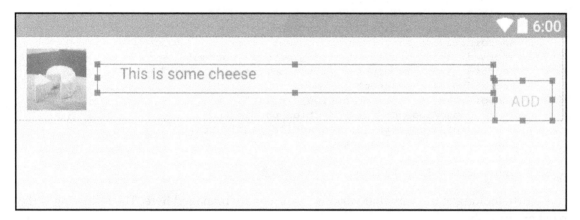

This may appear to be a very simple layout, but there is more to it than meets the eye. Here is the code for the three views:

The image:

```
<ImageView
    android:id="@+id/item_image"
    android:layout_width="@dimen/item_image_size"
    android:layout_height="@dimen/item_image_size"
    android:layout_gravity="center_vertical|end"
    android:layout_margin="@dimen/item_image_margin"
    android:scaleType="fitXY"
    android:src="@drawable/placeholder" />
```

The title:

```
<TextView
    android:id="@+id/item_name"
    android:layout_width="0dp"
    android:layout_height="wrap_content"
    android:layout_gravity="center_vertical"
    android:layout_weight="1"
    android:paddingBottom="@dimen/item_name_paddingBottom"
```

```
        android:paddingStart="@dimen/item_name_paddingStart"
        android:text="@string/placeholder"
        android:textSize="@dimen/item_name_textSize" />
```

The action button:

```
<Button
    android:id="@+id/action_add"
    style="?attr/borderlessButtonStyle"
    android:layout_width="wrap_content"
    android:layout_height="wrap_content"
    android:layout_gravity="center_vertical|bottom"
    android:layout_marginEnd="@dimen/action_marginEnd""
    android:minWidth="64dp"
    android:padding="@dimen/action_padding"
    android:paddingEnd="@dimen/action_paddingEnd"
    android:paddingStart="@dimen/action_paddingStart"
    android:text="@string/action_add_text"
    android:textColor="@color/colorAccent"
    android:textSize="@dimen/action_add_textSize" />
```

It is worth looking at the way that various resources are managed here. The following is the `dimens.xml` file:

```
<dimen name="item_name_paddingBottom">12dp</dimen>
<dimen name="item_name_paddingStart">24dp</dimen>
<dimen name="item_name_textSize">16sp</dimen>

<dimen name="item_image_size">64dp</dimen>
<dimen name="item_image_margin">12dp</dimen>

<dimen name="action_padding">12dp</dimen>
<dimen name="action_paddingStart">16dp</dimen>
<dimen name="action_paddingEnd">16dp</dimen>
<dimen name="action_marginEnd">12dp</dimen>
<dimen name="action_textSize">16sp</dimen>

<dimen name="fab_marginEnd">16dp</dimen>
```

It is immediately clear that several of these attributes carry the same values, and we could have achieved the same effect with only five. However, this can cause confusing code, especially when it comes to making changes later on, and despite this extravagant approach, there is still some hidden efficiency. The padding and margin settings for the action button will be the same for all such buttons across the app, as can be read clearly from their names and need only be declared once. Likewise, the text and image views in this layout are unique in this app and so are named accordingly. This also makes tweaking individual properties far clearer.

Finally, the use of `android:minWidth="64dp"` is a material stipulation intended to ensure all such buttons are wide enough for the average finger.

This completes the layout for this activity, and with our object factory in place as well, we can now populate our recycler view, as we did before, with a data adapter and a view holder.

Using the factory with the RecyclerView

As we saw briefly earlier in the book, RecyclerViews make use of an internal LayoutManager. This in turn communicates with the data set by use of an adapter. These adapters serve exactly the same function as the adapter design pattern we explored earlier in the book. The function may not appear so readily apparent, but it acts as a connection between a dataset and a recycler view's layout manager. The adapter crosses this bridge with its ViewHolder. The workings of the adapter are neatly separated from the client code, and all we need are a few lines to create a new adapter and layout manager.

With this in mind and our data ready, we can quickly put an adapter together by following these simple steps:

1. Begin by creating this new class in your main package:

```
public class DataAdapter extends
RecyclerView.Adapter<DataAdapter.ViewHolder> {
```

2. It requires the following field and constructor:

```
private List<Cheese> cheeses;

public DataAdapter(List<Cheese> cheeses) {
    this.cheeses = cheeses;
}
```

3. Now add the `ViewHolder` as an inner class, like so:

```
public static class ViewHolder extends RecyclerView.ViewHolder {
    public ImageView imageView;
    public TextView nameView;

    public ViewHolder(View itemView) {
        super(itemView);

        imageView = (ImageView) itemView.findViewById(R.id.item_image);
        nameView = (TextView) itemView.findViewById(R.id.item_name);
    }
}
```

4. There are three inherited methods that must be overridden. The `onCreateViewHolder()` method:

```
@Override
public DataAdapter.ViewHolder onCreateViewHolder(ViewGroup parent, int viewType) {
    Context context = parent.getContext();
    LayoutInflater inflater = LayoutInflater.from(context);

    View cheeseView = inflater.inflate(R.layout.item_view, parent, false);

    return new ViewHolder(cheeseView);
}
```

5. The `onBindViewHolder()` method:

```
@Override
public void onBindViewHolder(DataAdapter.ViewHolder viewHolder, int position) {
    Cheese cheese = cheeses.get(position);

    ImageView imageView = viewHolder.imageView;
    imageView.setImageResource(cheese.getImage());

    TextView nameView = viewHolder.nameView;
```

```
        nameView.setText(cheese.getName());
    }
```

6. The `getItemCount()` method:

```
@Override
public int getItemCount() {
    return cheeses.size();
}
```

That is the adapter now complete, and all we need to concern ourselves with is connecting it up to our data and recycler view. This we do from the `onCreate()` method of the main activity. First, we need to create a list of all our cheeses. With our pattern in place, this is remarkably simple. The following method can go anywhere but here is placed in the main activity:

```
private ArrayList<Cheese> buildList() {
    ArrayList<Cheese> cheeses = new ArrayList<>();

    cheeses.add(new Brie());
    cheeses.add(new Camembert());
    cheeses.add(new Cheddar());
    cheeses.add(new Emmental());
    cheeses.add(new Gouda());
    cheeses.add(new Manchego());
    cheeses.add(new Roquefort());

    return cheeses;
}
```

Note that you will need to import each of these classes from the Fillings package.

We can now connect this to our recycler view, by way of the adapter by adding these lines to the `onCreate()` method in the main activity:

```
RecyclerView recyclerView = (RecyclerView)
findViewById(R.id.recycler_view);

ArrayList<Cheese> cheeses = buildList();
DataAdapter adapter = new DataAdapter(cheeses);

recyclerView.setLayoutManager(new LinearLayoutManager(this));
recyclerView.setAdapter(adapter);

recyclerView.setHasFixedSize(true);
```

The first thing that stands out is just how little client code is required and how self-explanatory it is. Not only the code to set up the recycler view and adapter, but also the code to build the list. Without the pattern, we would have ended up with code like this:

```
cheeses.add(new Cheese("Emmental", R.drawable.emmental), 120, true, 65);
```

The project can now be tested on a device.

The linear layout manager that we used here is not the only one available to us. There are two other managers, one for grid layouts and one for staggered layouts. They can be applied like so:

```
recyclerView.setLayoutManager(new StaggeredGridLayoutManager(3,
StaggeredGridLayoutManager.VERTICAL));

recyclerView.setLayoutManager(new GridLayoutManager(this, 2));
```

This then only requires a little tweaking of the layout file and we can even provide alternative layouts and allow the user to select the one they prefer.

From a visual point of view, we have everything pretty much in place. However, with such a sparse item design, it might be nice to add dividers between items. This is not as straightforward as one might think, but it is nevertheless a simple and elegant process.

Adding dividers

Prior to the RecyclerView, the ListView came with its own divider element. The recycler view, on the other hand, does not. This should not be thought of as a shortfall, however, as this latter approach allows for more flexibility.

It may seem tempting to create a divider by adding a very narrow view at the bottom of the item layout, but this is considered very poor practice as when the item is moved or dismissed, the divider moves with it.

The RecyclerView uses an inner class, **ItemDecoration** to provide dividers between items, as well as spaces and highlights. It also has a very useful subclass, the ItemTouchHelper, which we will encounter shortly when we see how to swipe and dismiss cards.

First, follow these steps to add dividers to our recycler view:

1. Create a new ItemDecoration class:

```
public class ItemDivider extends RecyclerView.ItemDecoration
```

2. Include this Drawable field:

```
Private Drawable divider;
```

3. Followed by this constructor:

```
public ItemDivider(Context context) {
    final TypedArray styledAttributes =
context.obtainStyledAttributes(ATTRS);
    divider = styledAttributes.getDrawable(0);
    styledAttributes.recycle();
}
```

4. Then override the `onDraw()` method:

```
@Override
public void onDraw(Canvas canvas, RecyclerView parent, RecyclerView.State
state) {
    int left = parent.getPaddingLeft();
    int right = parent.getWidth() - parent.getPaddingRight();

    int count = parent.getChildCount();
    for (int i = 0; i < count; i++) {
        View child = parent.getChildAt(i);

        RecyclerView.LayoutParams params = (RecyclerView.LayoutParams)
child.getLayoutParams();

        int top = child.getBottom() + params.bottomMargin;
        int bottom = top + divider.getIntrinsicHeight();

        divider.setBounds(left, top, right, bottom);
        divider.draw(canvas);
    }
}
```

5. All that is needed now, is to instantiate the divider in the `onCreate()` method of the activity, after the `LayoutManager` has been set:

```
recyclerView.addItemDecoration(new ItemDivider(this));
```

This code provides the system divider between our items. The item decoration also makes it possible to create **custom dividers** very simply.

Just follow these two steps to see how it is done:

1. Create an XML file in the `drawable` directory called `item_divider.xml`, along these lines:

```
<?xml version="1.0" encoding="utf-8"?>
<shape xmlns:android="http://schemas.android.com/apk/res/android"
```

```
    android:shape="rectangle">
    <size android:height="1dp" />
    <solid android:color="@color/colorPrimaryDark" />
</shape>
```

2. Add a second constructor to the `ItemDivider` class, like this:

```
public ItemDivider(Context context, int resId) {
    divider = ContextCompat.getDrawable(context, resId);
}
```

3. Then replace the divider initialization in the activity, with this one:

```
recyclerView.addItemDecoration(new ItemDivider(this,
R.drawable.item_divider));
```

When run, these two techniques will produce results like those seen here:

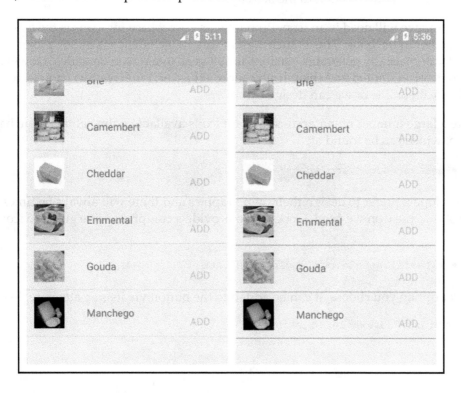

The preceding methods draw the divider before the view. If you have a fancy divider and wish parts of it to overlap the view, then you will need to override the `onDrawOver()` method instead, which will cause the divider to be drawn after the views.

It is now time to start to add a little functionality to our project. We will start by considering what functions we want to provide our floating action button.

Configuring the floating action button

So far, our layout provides only one action, the *add* action button on each list item. This will be used to include that filling in the user's eventual sandwich. It is always a good idea to ensure that the user is never more than one click away from spending their money, and so we will add a checkout function to the activity.

The first thing we will need is an icon. Probably the best source for icons is the asset studio we used earlier in the book. This is such a great way to include icons in our projects, mainly because it automatically generates versions for all available screen densities. However, the number of icons is limited and there is no checkout basket. We have two choices here: we can find an icon online or we can design our own.

There are a large number of material-compliant icons available online and Google have their own, which can be found at:

- `design.google.com/icons/`

Many developers prefer to design their own graphics and there will always be times when we cannot find the icon we need. Google also provide a comprehensive guide to icon design at:

- `material.google.com/style/icons.html`

Whichever option you choose, it can be added to the button via its `src` attribute, like so:

```
android:src="@drawable/ic_cart"
```

Having created our icon, we now need to consider color. According to material design guidelines, action and system icons should be the same color as either the primary or secondary text. These are not, as one might imagine, two shades of gray, rather they are defined by levels of transparency. This is done because it works far better on colored backgrounds than gray shades do. So far, we have used default text color and have not included this in our `styles.xml` file. This is easy enough to do given the rules regarding material text color are as follows:

Primary text on a *dark background* is 87% opaque black: **#DE000000**

Secondary text on a *dark background* is 54% opaque black: **#8A000000**

Primary text on a *light background* is 100% opaque white: **#FFFFFFFF**

Secondary text on a *light background* is 70% opaque white: **#B3FFFFFF**

To create primary and secondary text colors to our theme, add these lines to the `colors` file:

```
<color name="text_primary_dark">#DE000000</color>
<color name="text_secondary_dark">#8A000000</color>

<color name="text_primary_light">#FFFFFFFF</color>
<color name="text_secondary_light">#B3FFFFFF</color>
```

Then add the appropriate lines to the `styles` file, depending on the background shade, for example:

```
<item name="android:textColorPrimary">@color/text_primary_light</item>
<item name="android:textColorSecondary">@color/text_secondary_light</item>
```

If you have used an image asset or downloaded one of Google's material icons, then the system will automatically apply the primary text color to our FAB icon. Otherwise, you will need to color your icon directly.

We can now activate the toolbar and the FAB by following these two steps:

1. Add these lines to the main activity's `onCreate()` method:

```
Toolbar toolbar = (Toolbar) findViewById(R.id.toolbar);
setSupportActionBar(toolbar);
```

2. Add the following click listener to the `onCreate()` method of its activity:

```
FloatingActionButton fab = (FloatingActionButton) findViewById(R.id.fab);
fab.setOnClickListener(new View.OnClickListener() {

    @Override
    public void onClick(View view) {
        // SYSTEM DISMISSES DIALOG
    }
});
```

The FAB icon and toolbar title will now be visible and animate correctly when the view is scrolled:

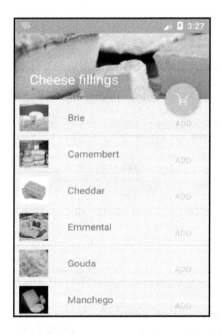

Clicking on the FAB should take the user to another activity, the checkout activity. However, they may have clicked the button in error, and therefore we should first present them with a dialog for them to confirm the selection.

The dialog builder

As well as being essential to all but a few apps, Android dialogs offer a great way to see how the framework itself employs design patterns. In this case, it is the dialog builder, which strings together a series of setters to build our dialog.

In the current situation, all we really need is a very simple dialog allowing the user to confirm their selection, but dialog construction is a very interesting topic and so we will take a closer look at how it is done and how inbuilt builder patterns are used to construct them.

The dialog we are about to build will, if confirmed, take the user to another activity, so before we do that we should create that activity. This is easily done by selecting New | Activity | Blank Activity from the project explorer menu. Here we have called it CheckoutActivity.java.

Once you have created this activity, follow these two steps:

1. The floating action button onClickListener will build and inflate our dialog. It is quite lengthy, so create a new method called buildDialog(): and add the following two lines to the bottom of the onCreate() method:

```
fab = (FloatingActionButton) findViewById(id.fab);
buildDialog(fab);
```

2. Then define the method like this:

```
private void buildDialog(FloatingActionButton fab) {
    fab.setOnClickListener(new View.OnClickListener() {

        @Override
        public void onClick(View view) {
            AlertDialog.Builder builder = new
AlertDialog.Builder(MainActivity.this);

            LayoutInflater inflater =
MainActivity.this.getLayoutInflater();

        builder.setTitle(R.string.checkout_dialog_title)

            .setMessage(R.string.checkout_dialog_message)

            .setIcon(R.drawable.ic_sandwich_primary)

            .setPositiveButton(R.string.action_ok_text, new
```

```
DialogInterface.OnClickListener() {

                public void onClick(DialogInterface dialog, int id) {
                    Intent intent = new Intent(MainActivity.this,
CheckoutActivity.class);
                    startActivity(intent);
                }
            })

            .setNegativeButton(R.string.action_cancel_text, new
DialogInterface.OnClickListener() {

                public void onClick(DialogInterface dialog, int id) {
                    // SYSTEM DISMISSES DIALOG
                }
            });

        AlertDialog dialog = builder.create();
        dialog.show();
    }
  });
}
```

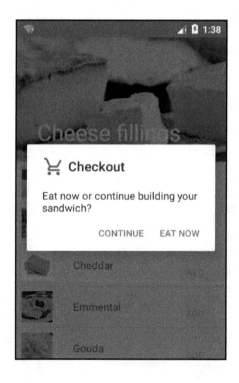

For such a simple dialog, it is unnecessary to have a title and an icon and these are included by way of example only. There are many other attributes provided by the `AlertDialog.Builder` and a comprehensive guide can be found at:

developer.android.com/reference/android/app/AlertDialog.Builder.html

This provides a convenient way to put together almost any alert dialog we can think of, but it has some shortfalls. For example, the above dialog uses the default theme to color the button text. With our customized theme, it would be nice to see this applied to our dialogs as well. This is easily achieved by creating a custom dialog.

Custom dialogs

As you would expect, a custom dialog is defined with an XML layout file in the same way we might design any other layout. Furthermore, we can inflate this layout during our builder chain, which means we can combine custom and default features in the same dialog.

There are just two steps to customize our dialog:

1. Firstly, create a new layout resource file called `checkout_dialog.xml` and complete it like so:

```xml
<?xml version="1.0" encoding="utf-8"?>
<LinearLayout xmlns:android="http://schemas.android.com/apk/res/android"
    android:layout_width="match_parent"
    android:layout_height="match_parent"
    android:orientation="vertical"
    android:theme="@style/AppTheme">

    <ImageView
        android:id="@+id/dialog_title"
        android:layout_width="match_parent"
        android:layout_height="@dimen/dialog_title_height"
        android:src="@drawable/dialog_title" />

    <TextView
        android:id="@+id/dialog_content"
        android:layout_width="wrap_content"
        android:layout_height="wrap_content"
        android:paddingStart="@dimen/dialog_message_padding"
        android:text="@string/checkout_dialog_message"
        android:textAppearance="?android:attr/textAppearanceSmall"
        android:textColor="@color/text_secondary_dark" />
```

```
</LinearLayout>
```

2. Then, edit the `buildDialog()` method to match the one seen here. The changes from the previous method have been highlighted:

```
private void buildDialog(FloatingActionButton fab) {
    fab.setOnClickListener(new View.OnClickListener() {

        @Override
        public void onClick(View view) {
            AlertDialog.Builder builder = new
AlertDialog.Builder(MainActivity.this);

            LayoutInflater inflater =
MainActivity.this.getLayoutInflater();

            builder.setView(inflater.inflate(layout.checkout_dialog, null))

                .setPositiveButton(string.action_ok_text, new
DialogInterface.OnClickListener() {
                    public void onClick(DialogInterface dialog, int id)
{
                        Intent intent = new Intent(MainActivity.this,
CheckoutActivity.class);
                        startActivity(intent);
                    }
                })
                .setNegativeButton(string.action_cancel_text, new
DialogInterface.OnClickListener() {
                    public void onClick(DialogInterface dialog, int id)
{
                        // System dismisses dialog
                    }
                });

            AlertDialog dialog = builder.create();
            dialog.show();

            Button cancelButton =
dialog.getButton(DialogInterface.BUTTON_NEGATIVE);
cancelButton.setTextColor(getResources().getColor(color.colorAccent));

            Button okButton =
dialog.getButton(DialogInterface.BUTTON_POSITIVE);
okButton.setTextColor(getResources().getColor(color.colorAccent));
        }
    });
}
```

Here, we used the `AlertDialog.Builder` to set the view to our custom layout. This requires the layout resource and the parent, but in this case, we are building from within the listener, so it remains `null`.

When tested on a device, the output should resemble the following screenshot:

It is worth noting that when defining string resources for buttons, it is considered better practice to *not* capitalize the whole string, but only the first letter. For example, the following definitions created the text on the buttons in the previous example:

```
<string name="action_ok_text">Eat now</string>
<string name="action_cancel_text">Continue</string>
```

In this example, we customized the title and content of the dialog, but still used the provided OK and CANCEL buttons, and we can mix and match our own customizations with many of the dialog's setters.

Before we move on, we will provide one more form of functionality to the recycler view, swipe and dismiss behavior.

Adding swipe and dismiss actions

It is unlikely that we would need swipe and dismiss behavior in this particular app, as the lists are short and there is little to be gained by allowing users to edit them. However, so that we can see how this important and useful function is applied, we will implement it here even though we won't be including it in the final design.

Swiping, as well as dragging and dropping, is largely managed by the **ItemTouchHelper**, which is a type of RecyclerView.ItemDecoration. The callbacks provided for this class allow us to detect item movement and direction and to intercept these actions and respond to them in code.

As you can see here, there are just a few steps to implementing swipe and dismiss behavior:

1. Firstly, our list is now going to change length, so remove the line `recyclerView.setHasFixedSize(true);` or set it to `false`.

2. It is always a good idea to keep our `onCreate()` methods as simple as possible, as there can often be a great deal going on there. We will create a separate method to initialize our item touch helper and call it from `onCreate()`. Here is the method:

```
private void initItemTouchHelper() {
    ItemTouchHelper.SimpleCallback callback = new
ItemTouchHelper.SimpleCallback(0, ItemTouchHelper.LEFT |
ItemTouchHelper.RIGHT) {

        @Override
        public boolean onMove(RecyclerView recyclerView,
RecyclerView.ViewHolder viewHolder, RecyclerView.ViewHolder viewHolder1) {
            return false;
        }

        @Override
        public void onSwiped(RecyclerView.ViewHolder viewHolder, int
direction) {
            int position = viewHolder.getAdapterPosition();
            adapter.removeItem(position);
        }
    };

    ItemTouchHelper itemTouchHelper = new ItemTouchHelper(callback);
    itemTouchHelper.attachToRecyclerView(recyclerView);
}
```

3. Now add the following line to the onCreate() method:

```
InitItemTouchHelper();
```

Despite performing half a dozen functions, the onCreate() method still remains short and clear:

```
@Override
protected void onCreate(Bundle savedInstanceState) {
    super.onCreate(savedInstanceState);
    setContentView(layout.activity_main);

    Toolbar toolbar = (Toolbar) findViewById(id.toolbar);
    setSupportActionBar(toolbar);

    final ArrayList<Cheese> cheeses = buildList();
    adapter = new DataAdapter(cheeses);

    recyclerView = (RecyclerView) findViewById(id.recycler_view);
    recyclerView.setLayoutManager(new LinearLayoutManager(this));
    recyclerView.addItemDecoration(new ItemDivider(this));
    recyclerView.setAdapter(adapter);

    initItemTouchHelper();

    fab = (FloatingActionButton) findViewById(id.fab);
    buildDialog(fab);
}
```

If you test the app at this point, you will notice that although items disappear from the screen when swiped, the gap does not close up. This is because we have not yet informed the recycler view that it has been removed. Although this can be done from the initItemTouchHelper() method, it really belongs in the adapter class, as it utilizes its methods. Add the following method to the adapter to complete this task:

```
public void removeItem(int position) {
    cheeses.remove(position);
    notifyItemRemoved(position);
    notifyItemRangeChanged(position, cheeses.size());
```

The recycler view list will now reorder when an item is removed:

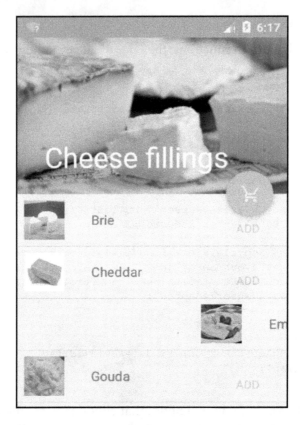

In this example, the user can swipe an item either way to dismiss it, and this is fine for our purposes here, but there are many times when this distinction is very useful. Many mobile applications use swipe right to accept an item and swipe left to dismiss it. This is easily implemented by using the onSwiped() method's direction parameter. For example:

```
if (direction == ItemTouchHelper.LEFT) {
    Log.d(DEBUG_TAG, "Swiped LEFT");
} else {
    Log.d(DEBUG_TAG, "Swiped RIGHT");
}
```

Earlier in the chapter, we used a native pattern, the AlertDialog.Builder, to construct a layout. As is meant to be the case with creational patterns, the logic behind the processes was hidden from us, but the builder design pattern provides a very good mechanism for constructing layouts and view groups from individual view components, as we shall see next.

Constructing layout builders

So far in this book, all the layouts we have constructed have been static XML definitions. As you would expect, however, it is perfectly possible to construct and inflate UIs dynamically from our source code. Furthermore, Android layouts lend themselves very nicely to the builder pattern, as we saw with our alert dialog, as they are comprised of an ordered collection of smaller objects.

The following example will follow the builder design pattern to inflate a linear layout from a series of predefined *layout views*. As before, we will build up from interfaces to abstractions and concrete classes. We will create two kinds of layout item, a title or *headline* view and a *content* view. We then make several concrete examples of these that can then be constructed by the builder. As there are some features that all views have in common (text and background colors in this case), we will avoid having to duplicate methods by having another interface, with its own concrete extensions to handle this shading.

To best see how this works, start a new Android project and follow these steps to construct the model:

1. Create an inner package called `builder`. Add all the following classes to this package.
2. Create the following interface for our view classes:

```
public interface LayoutView {

    ViewGroup.LayoutParams layoutParams();
    int textSize();
    int content();
    Shading shading();
    int[] padding();
}
```

3. Now create the interface for the text and background colors, like so:

```
public interface Shading {

    int shade();
    int background();
}
```

4. We will create the concrete examples of `Shading`. They look like this:

```
public class HeaderShading implements Shading{

    @Override
    public int shade() {
        return R.color.text_primary_dark;
    }

    @Override
    public int background() {
        return R.color.title_background;
    }
}


public class ContentShading implements Shading{

    ...
        return R.color.text_secondary_dark;
    ...


    ...
        return R.color.content_background;
    ...
}
```

5. Now we can create abstract implementations of the two types of view we want. These should match the following:

```
public abstract class Header implements LayoutView {

    @Override
    public Shading shading() {
        return new HeaderShading();
    }
}


public abstract class Content implements LayoutView {

    ...
        return new ContentShading();
    ...
}
```

6. Next, we need to create concrete classes of both these types. First the headers:

```java
public class Headline extends Header {

    @Override
    public ViewGroup.LayoutParams layoutParams() {
        final int width = ViewGroup.LayoutParams.MATCH_PARENT;
        final int height = ViewGroup.LayoutParams.WRAP_CONTENT;

        return new ViewGroup.LayoutParams(width,height);
    }

    @Override
    public int textSize() {
        return 24;
    }

    @Override
    public int content() {
        return R.string.headline;
    }

    @Override
    public int[] padding() {
        return new int[]{24, 16, 16, 0};
    }
}


public class SubHeadline extends Header {

    ...

    @Override
    public int textSize() {
        return 18;
    }

    @Override
    public int content() {
        return R.string.sub_head;
    }

    @Override
    public int[] padding() {
        return new int[]{32, 0, 16, 8};
    }
    ...
```

7. Then Content:

```java
public class SimpleContent extends Content {

    @Override
    public ViewGroup.LayoutParams layoutParams() {
        final int width = ViewGroup.LayoutParams.MATCH_PARENT;
        final int height = ViewGroup.LayoutParams.MATCH_PARENT;

        return new ViewGroup.LayoutParams(width, height);
    }

    @Override
    public int textSize() {
        return 14;
    }

    @Override
    public int content() {
        return R.string.short_text;
    }

    @Override
    public int[] padding() {
        return new int[]{16, 18, 16, 16};
    }
}


public class DetailedContent extends Content {

    ...
        final int height = ViewGroup.LayoutParams.WRAP_CONTENT;
    ...
    @Override
    public int textSize() {
        return 12;
    }

    @Override
    public int content() {
        return R.string.long_text;
    }

    ...
```

This completes our model. We have two individual views and color settings for each type of view. We can now create a helper class to put these views together in whichever order we wish. Here we will have just two, one for a simple output and one for a more detailed layout.

This is how the builder looks:

```
public class LayoutBuilder {

    public List<LayoutView> displayDetailed() {
        List<LayoutView> views = new ArrayList<LayoutView>();
        views.add(new Headline());
        views.add(new SubHeadline());
        views.add(new DetailedContent());
        return views;
    }

    public List<LayoutView> displaySimple() {
        List<LayoutView> views = new ArrayList<LayoutView>();
        views.add(new Headline());
        views.add(new SimpleContent());
        return views;
    }
}
```

The class diagram for this pattern is as follows:

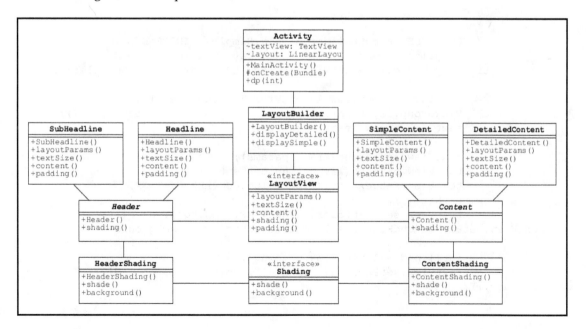

As is the intention with the builder pattern and other patterns in general, all the work we have just gone to serves to hide the model logic from the client code, in our case the current activity and the `onCreate()` method in particular.

We could of course inflate these views in the default root view group as provided by the main XML activity, but it is often useful to generate these dynamically too, especially if we want to generate nested layouts.

The following activity demonstrates how we can now use the builder to inflate layouts dynamically:

```
public class MainActivity extends AppCompatActivity {
    TextView textView;
    LinearLayout layout;

    @Override
    protected void onCreate(Bundle savedInstanceState) {
        final int width = ViewGroup.LayoutParams.MATCH_PARENT;
        final int height = ViewGroup.LayoutParams.WRAP_CONTENT;

        super.onCreate(savedInstanceState);

        layout = new LinearLayout(this);
        layout.setOrientation(LinearLayout.VERTICAL);
        layout.setLayoutParams(new ViewGroup.LayoutParams(width, height));

        setContentView(layout);

        // COULD USE layoutBuilder.displaySimple() INSTEAD
        LayoutBuilder layoutBuilder = new LayoutBuilder();
        List<LayoutView> layoutViews = layoutBuilder.displayDetailed();

            for (LayoutView layoutView : layoutViews) {
            ViewGroup.LayoutParams params = layoutView.layoutParams();
            textView = new TextView(this);

            textView.setLayoutParams(params);
            textView.setText(layoutView.content());
            textView.setTextSize(TypedValue.COMPLEX_UNIT_SP,
layoutView.textSize());
            textView.setTextColor(layoutView.shading().shade());
textView.setBackgroundResource(layoutView.shading().background());

            int[] pad = layoutView.padding();
            textView.setPadding(dp(pad[0]), dp(pad[1]), dp(pad[2]),
dp(pad[3]));
```

```
            layout.addView(textView);
        }
    }
}
```

You will also need the following method, which is used to convert from px to dp:

```
public int dp(int px) {
    final float scale = getResources().getDisplayMetrics().density;
    return (int) (px * scale + 0.5f);
}
```

Which when run on a device will produce one of the following two UIs:

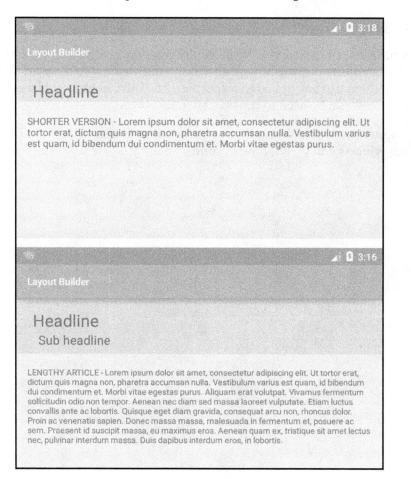

As expected, the client code is simple, short, and easy to follow.

It is not necessary to use either programmatic or static layouts, and the two can be mixed. Views can be designed in XML and then inflate them the way we did here in Java. We can even keep the same pattern we used here.

There is a lot more that could be covered here, such as how to include other types of views, such as images using adapter or bridge patterns, but we will cover combining patterns later in the book. For now, we have seen how a layout builder works in principle and how it separates its logic from the client code.

Summary

This chapter has covered quite a lot. We began by creating a collapsing toolbar and a functional recycler view. We saw how to add basic functionality to much of our layout, and how a factory pattern can be applied to a specific case. This led us to explore how builders, internal and created, can be used to construct detailed layouts.

In the next chapter, we will look further into responding to user activity and, now we have some working widgets and views, how to connect them to some useful logic.

7

Combining Patterns

We have seen how patterns can help us organize our code and how this can be applied specifically to Android apps, but we have only applied one pattern at a time. As the tasks we need to perform become more complex, we will need to apply several patterns, such as decorators and builders, at once and even combine them into **hybrid patterns**, and this is what we will do in this chapter.

We will begin by considering a more complex user interface (UI) and the code behind it. This will require us to think a little more precisely about what we actually want our application to do. This in turn will lead us to look at the **prototype pattern**, which provides a very efficient method for creating objects from an original, clone, object.

The **decorator pattern** is explored next, and we see how it can be used to add extra functionality to existing classes. Often referred to as a wrapper, the decorator is used to provide additional functionality to existing code. This is particularly useful to our sandwich builder app as it allows us to include options such as ordering an open sandwich or having the bread toasted. These are not in themselves ingredients, but nevertheless something a sandwich vendor would wish to provide. The decorator pattern is ideal for this task.

Having looked briefly at the alternatives, we construct a builder pattern to form the basis of our system, connecting it to a UI so that a simple sandwich, with a choice of options and ingredients, can be put together by the user. We then connect a decorator to this builder to provide further options.

In this chapter you will learn how to do the following:

- Create a prototype pattern
- Create a decorator pattern
- Extend a decorator
- Connect a builder to a UI

- Manage compound buttons
- Combine patterns

We are now in a position to start thinking more about the details of our app and what it can and should do. We need to think about the potential customer and design something that is simple and pleasant to use. Features need to be easily accessed and obvious in their nature and most of all, they need to be able to construct their desired sandwich with a minimum number of clicks. Later on, we will see how users can store favorites and how we can provide partially built sandwiches for the user to customize rather than build from the ground up. For now, we will take a look at how to classify our sandwich-related objects and classes.

Outlining specifications

In the previous chapter, we created a simple list of sandwich ingredient objects using a factory pattern and connected it to a layout. However, we only represented a single type of filling. Before we can create a more sophisticated system, we need to plan our data structure, and to do that we need to consider the choices we present the user.

Firstly, what options can we offer the user to make the process simple, fun, and intuitive? Here is a list of functions a potential user may want from such an app:

- Order an off-the-shelf sandwich, with no customization
- Customize an off-the-shelf sandwich
- Start with some basic ingredients and build from there
- Order or customize a sandwich they have had before
- Build a sandwich from scratch
- Review and edit their sandwich at any time

Previously, we created an individual menu for cheeses, but a category for each food type may offer a clumsy solution: a user wanting a bacon, lettuce, and tomato sandwich may have to visit three separate menus. There are many different ways we could solve this problem, and it is largely a matter of personal choice. Here, we will try to follow the course we might take when making a sandwich for ourselves, which could be described by the following list:

1. Bread
2. Butter
3. Fillings

4. Toppings

By toppings, I mean mayonnaise, pepper, mustard, and so on. We will use these categories as the basis for our class structure. It would be nice if they could all belong to the same class type, but there are one or two subtle differences that forbid that:

Bread: No one is going to order a sandwich without bread; it wouldn't be a sandwich, and we would be forgiven for thinking that it could be treated like any other ingredient. However, we are going to offer the choice of an open sandwich, and to complicate things for ourselves, the option of having it toasted.

Butter: Again, it would be easy to think that adding butter goes without saying, but some customers will want a low-fat spread, or even none at all. Fortunately, there is a pattern that suits this purpose very well: the decorator pattern.

Fillings and toppings: Although these classes could very easily share identical properties and instances if both extend from the same class, we will treat them separately as this will make constructing menus far clearer.

With these specifications in place, we can start to think about how the top-level menu will look. We will use a sliding drawer navigation view and offer the following options:

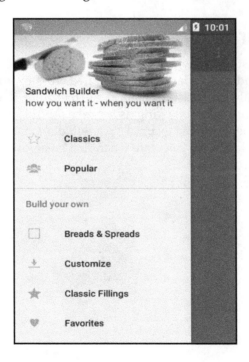

This gives us a rough idea of what we are aiming at. One of the advantages of using patterns is that the ease at which they can be modified means we can take a more intuitive approach to development, safe in the knowledge that even large-scale changes often only require editing a minimum of code.

Our next step is to select a suitable pattern for the outlined task. We are familiar with both factories and builders and how they could both accomplish what we want, but there is another creational pattern, the prototype, which is also very handy, and although we will not be using it in this situation there will be times when we might, and certainly times when you will.

The prototype pattern

The prototype design pattern performs similar tasks to other creational patterns, such as builders and factories, but it takes a very different approach. Rather than rely heavily on a number of hard-coded sub-classes, the prototype, as its name suggests, makes copies from an original, vastly reducing the number of sub-classes required and any lengthy creation processes.

Setting up a prototype

The prototype is at its most useful when the creation of an instance is expensive in some way. This could be the loading of a large file, a detailed cross-examination of a database, or some other computationally expensive operation. Furthermore, it allows us to decouple cloned objects from their originals, allowing us to make modifications without having to re-instantiate each time. In the following example, we will demonstrate this using functions that take some considerable time to calculate when first created: the nth prime number and the nth Fibonacci number.

Viewed diagrammatically, our prototype will look like this:

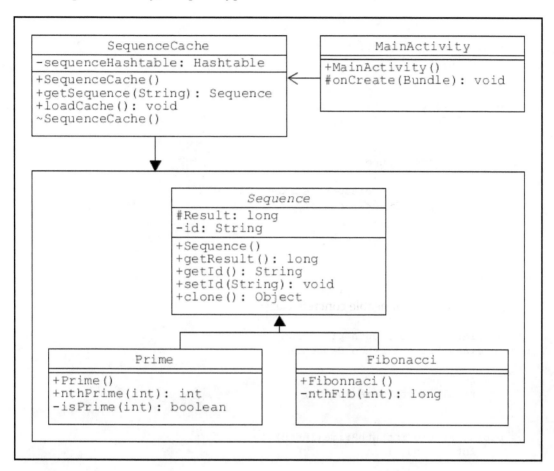

We will not need the prototype pattern in our main app as there are very few expensive creations. However, it is vitally important in many situations and should not be neglected. Follow these steps to apply a prototype pattern:

1. We will start with the the following abstract class:

```
public abstract class Sequence implements Cloneable {
    protected long result;
    private String id;

    public long getResult() {
        return result;
    }
}
```

```java
    public String getId() {
        return id;
    }

    public void setId(String id) {
        this.id = id;
    }

    public Object clone() {
        Object clone = null;

        try {
            clone = super.clone();

        } catch (CloneNotSupportedException e) {
            e.printStackTrace();
        }

        return clone;
    }
}
```

2. Next, add this cloneable concrete class:

```java
// Calculates the 10,000th prime number
public class Prime extends Sequence {

    public Prime() {
        result = nthPrime(10000);
    }

    public static int nthPrime(int n) {
        int i, count;

        for (i = 2, count = 0; count < n; ++i) {
            if (isPrime(i)) {
                ++count;
            }
        }

        return i - 1;
    }

    // Test for prime number
    private static boolean isPrime(int n) {

        for (int i = 2; i < n; ++i) {
            if (n % i == 0) {
```

```
                    return false;
            }
        }

        return true;
    }
}
```

3. Add another `Sequence` class, for the Fibonacci numbers, like so:

```java
// Calculates the 100th Fibonacci number
public class Fibonacci extends Sequence {

    public Fibonacci() {
        result = nthFib(100);
    }

    private static long nthFib(int n) {
        long f = 0;
        long g = 1;

        for (int i = 1; i <= n; i++) {
            f = f + g;
            g = f - g;
        }

        return f;
    }
}
```

4. Next, create the cache class, like this:

```java
public class SequenceCache {
    private static Hashtable<String, Sequence> sequenceHashtable = new
Hashtable<String, Sequence>();

    public static Sequence getSequence(String sequenceId) {

        Sequence cachedSequence = sequenceHashtable.get(sequenceId);
        return (Sequence) cachedSequence.clone();
    }

        public static void loadCache() {

        Prime prime = new Prime();
        prime.setId("1");
        sequenceHashtable.put(prime.getId(), prime);
```

```
            Fibonacci fib = new Fibonacci();
            fib.setId("2");
            sequenceHashtable.put(fib.getId(), fib);
      }
   }
```

5. Add three TextViews to your layout, and then add the code to your MainActivity's `onCreate()` method.

6. Add these lines to the client code:

```
// Load the cache once only
SequenceCache.loadCache();

// Lengthy calculation and display of prime result
Sequence prime = (Sequence) SequenceCache.getSequence("1");
primeText.setText(new StringBuilder()
        .append(getString(R.string.prime_text))
        .append(prime.getResult())
        .toString());

// Lengthy calculation and display of Fibonacci result
SSequence fib = (Sequence) SequenceCache.getSequence("2");
fibText.setText(new StringBuilder()
        .append(getString(R.string.fib_text))
        .append(fib.getResult())
        .toString());
```

As you can see, the preceding code creates the pattern but does not demonstrate it. Once loaded, the cache can create instant copies of our previously expensive output. Furthermore, we can modify the copy, making the prototype very useful when we have a complex object and want to modify just one or two properties.

Applying the prototype

Consider a detailed user profile such as you might find on a social media site. Users modify details such as images and text, but the overall structure is the same for all profiles, making it an ideal candidate for a prototype pattern.

To put this principle into practice, include the following code in your client source code:

```
// Create a clone of already constructed object
Sequence clone = (Fibonacci) new Fibonacci().clone();

// Modify the resultlong result = clone.getResult() / 2;
```

```
// Display the result quickly
cloneText.setText(new StringBuilder()
.append(getString(R.string.clone_text))          .append(result)
.toString());
```

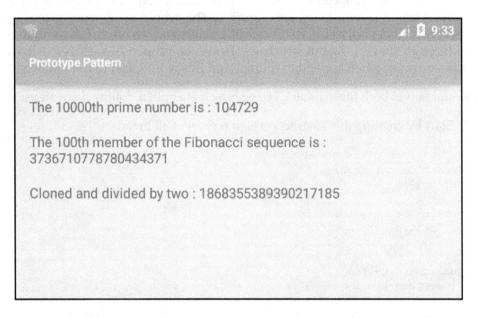

The prototype is a very useful pattern in many occasions where we have expensive objects to create or when we face a proliferation of sub-classes. However, this is not the only pattern that helps reduce excessive sub-classing, and this leads us on to another design pattern: the **decorator**.

The decorator design pattern

Regardless of object creation expense, there are still times when the nature of our model will necessitate an unreasonable number of sub-classes, and this is where the decorator comes in extremely handy.

Take the bread in our sandwich app, for example. We would like to offer several types of bread, but in addition, we want to offer the choice of having the bread toasted, the sandwich open, and a selection of spreads. By creating toasted and open versions for each bread type, the project would very soon become unmanageable. The decorator allows us to add functionality and properties to an object during runtime without having to make any changes to the original class structure.

Setting up a decorator

One might think that properties such as *toasted* and *open* could be included as part of the *bread* class, but this itself can lead to increasingly unwieldy code. Say that we want *bread* and *filling* to inherit from the same class, say *ingredient*. This would make sense as they have properties in common, such as price and calorific value, and we want them both to be displayed through the same layout structures. However, properties such as toasted and spread make no sense when applied to fillings, and this would lead to redundancy.

The decorator solves both these issues. To see how it is applied, follow these steps:

1. Start by creating this abstract class to represent all breads:

```
public abstract class Bread {
    String description;
    int kcal;

    public String getDescription() {
        return description;
    }

    public int getKcal() {
        return kcal;
    }
}
```

2. Next, create concrete instances, like so:

```
public class Bagel extends Bread {

    public Bagel() {
        description = "Bagel";
        kcal = 250;
    }
}


public class Bun extends Bread {

    public Bun() {
        description = "Bun";
        kcal = 150;
    }
}
```

3. Now we need an abstract decorator. It looks like this:

```
// All bread treatments extend from this
public abstract class BreadDecorator extends Bread {

    public abstract String getDescription();

    public abstract int getKcal();
}
```

4. We need four extensions of this decorator to represent two types of spread and both open and toasted sandwiches. First, the `Butter` decorator:

```
public class Butter extends BreadDecorator {
    private Bread bread;

public Butter(Bread bread) {
        this.bread = bread;
    }

    @Override
    public String getDescription() {
        return bread.getDescription() + " Butter";
    }

    @Override
    public int getKcal() {
        return bread.getKcal() + 50;
    }
}
```

5. Only the values returned by the getters differ in the other three classes. They are as follows:

```
public class LowFatSpread extends BreadDecorator {
        return bread.getDescription() + " Low fat spread";
        return bread.getKcal() + 25;
}

public class Toasted extends BreadDecorator {
        return bread.getDescription() + " Toasted";
        return bread.getKcal() + 0;
}

public class Open extends BreadDecorator {
        return bread.getDescription() + " Open";
        return bread.getKcal() / 2;
```

```
    }
```

That is all that is required to set up our decorator pattern. All we need to do now is connect it to a working interface of some sort. Later, we will use a menu for selecting the bread and then a dialog to add the *decoration*.

Applying the decorator

The user will have to choose between butter and low-fat spread (although a *no spread* option could be included by adding another decorator), but can choose to have their sandwich both toasted and open.

For now, we will use the debugger to test various combinations by adding lines such as the following to the governing activity's `onCreate()` method. Note how the objects are chained:

```
Bread bagel = new Bagel();

LowFatSpread spread = new LowFatSpread(bagel);

Toasted toast = new Toasted(spread);

Open open = new Open(toast);

Log.d(DEBUG_TAG, open.getDescription() + " " + open.getKcal());
```

This should produce outputs like these:

```
D/tag: Bagel Low fat spread 275
D/tag: Bun Butter Toasted 200
D/tag: Bagel Low fat spread Toasted Open 137
```

Diagrammatically, our decorator pattern can be expressed like so:

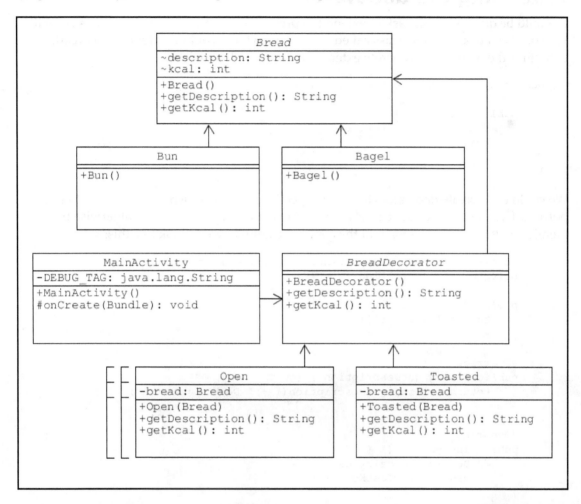

The decorator design pattern is an extremely useful development tool and can be applied to a multitude of situations. As well as helping us maintain a manageable number of concrete classes, we could also have our bread super class inherit from the same interface as the filling class and still behave differently.

Extending the decorator

It would be quite a simple task to extend the preceding pattern to cater for fillings as well. We could create an abstract class called `Fillings`, which would be identical to Bread, except for the name, with concrete extensions along the lines of this:

```java
public class Lettuce extends Filling {

    public Lettuce() {
        description = "Lettuce";
        kcal = 1;
    }

}
```

We could even create decorators that were specific for fillings such as ordering a double portion. The `FillingDecorator` class would extend from `Filling` but otherwise be identical to `BreadDecorator`, and the concrete example would look like this:

```java
public class DoublePortion extends FillingDecorator {
    private Filling filling;

    public DoublePortion(Filling filling) {
        this.filling = filling;
    }

    @Override
    public String getDescription() {
        return filling.getDescription() + " Double portion";
    }

    @Override
    public int getKcal() {
        // Double the calories
        return filling.getKcal() * 2;
    }
}
```

The way that we chained our decorators together to produce a compound string is very similar to the way a builder works, and we could indeed use this pattern to generate an entire sandwich along with all its trimmings. However, as is often the case, there is more than one candidate for this task. As we saw earlier in the book, builders and abstract factories are both capable of producing complex objects. Before we decide on our model, we need to find the most suitable pattern or, better still, a combination of patterns.

The builder pattern would seem the most obvious choice, so we will take a look at that first.

A sandwich builder pattern

The builder pattern is purpose-built for combining simple objects to form one complex object, and this forms a perfect analogy of making a sandwich. We encountered a generalized builder pattern earlier in the book, but now we need to adapt it for a specific function. Furthermore, we will be connecting the pattern to a working UI so that a sandwich can be constructed according to user selections rather than the set meal demonstrated in previous builder examples.

Applying the pattern

To keep the code short and simple, we will only create two concrete classes of each ingredient type, and we will use buttons and a text view to display the output rather than a recycler view. Simply follow these steps to create our sandwich builder pattern:

1. Begin with the following interface:

```
public interface Ingredient {

    public String description();

    public int kcal();
}
```

2. Create these two abstract implementations of Ingredient. They are empty for now, but we will need them later:

```
public abstract class Bread implements Ingredient {

    // Base class for all bread types
}


public abstract class Filling implements Ingredient {

    // Base class for all possible fillings
}
```

3. We will need just two concrete examples of each ingredient type. Here is one, the Bagel class:

```
public class Bagel extends Bread {

    @Override
    public String description() {
        return "Bagel";
    }

    @Override
    public int kcal() {
        return 250;
    }
}
```

4. Create another Bread called Bun and two Filling classes called Egg and Cress.
5. Provide these classes with any description and calorific values you like.
6. Now we can create the sandwich class itself, like so:

```
public class Sandwich {
    private List<Ingredient> ingredients = new ArrayList<Ingredient>();

    // Add individual ingredients
    public void addIngredient(Ingredient i) {
        ingredients.add(i);
    }

    // Calculate total calories
    public int getKcal() {
        int kcal = 0;

        for (Ingredient ingredient : ingredients) {
            kcal += ingredient.kcal();
        }

        return kcal;
    }

    // Return all ingredients when selection is complete
    public String getSandwich() {
        String sandwich = "";

        for (Ingredient ingredient : ingredients) {
            sandwich += ingredient.description() + "\n";
        }
```

```
            return sandwich;
    }
}
```

7. The sandwich builder class does not build set meals, as in previous examples, but is used to add ingredients as requested. It is as follows:

```
public class SandwichBuilder {

    public Sandwich build(Sandwich sandwich, Ingredient ingredient) {
        sandwich.addIngredient(ingredient);
        return sandwich;
    }
}
```

This completes the pattern itself, but before we move on to create the UI, we need to address the empty abstract classes Bread and Filling. They appear to be utterly superfluous but there are two reasons why we have done this.

Firstly, by defining their methods, description() and kcal(), in a common interface we can more easily create ingredients that are neither filling or bread by implementing the interface itself.

To see how, add the following class to the project:

```
public class Salt implements Ingredient {

    @Override
    public String description() {
        return "Salt";
    }

    @Override
    public int kcal() {
        return 0;
    }
}
```

This gives us the following class structure:

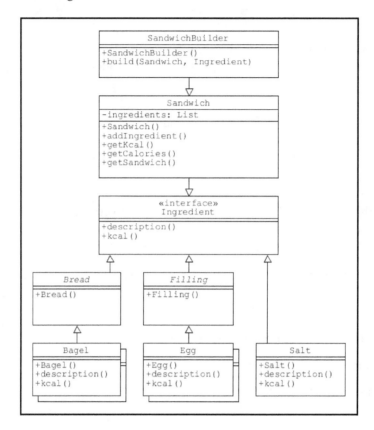

The second reason for including these abstract classes is more interesting. The `BreadDecorator` class in the previous example worked directly with the abstract `Bread` class and by maintaining that structure we can easily connect decorators to our ingredient types. We will move on to this shortly, but first we are going to build a UI to run our sandwich builder on.

Connecting to a UI

In this demonstration, we have two types of filling and two breads. They can select as many or as few fillings as they wish, but only one type of bread, which makes the selection a good candidate for the use of **check boxes** and **radio buttons** respectively. There is also an option to add salt, and this is the kind of binary choice that perfectly suits the **switch widget**.

To begin with we need a layout. Here are the steps required:

1. Begin with a vertical linear layout.
2. Then include the radio button group, like so:

```
<RadioGroup xmlns:android="http://schemas.android.com/apk/res/android"
    android:layout_width="fill_parent"
    android:layout_height="wrap_content"
    android:orientation="vertical">

    <RadioButton
        android:id="@+id/radio_bagel"
        android:layout_width="wrap_content"
        android:layout_height="wrap_content"
        android:checked="false"
        android:paddingBottom="@dimen/padding"
        android:text="@string/bagel" />

    <RadioButton
        android:id="@+id/radio_bun"
        android:layout_width="wrap_content"
        android:layout_height="wrap_content"
        android:checked="true"
        android:paddingBottom="@dimen/padding"
        android:text="@string/bun" />
</RadioGroup>
```

3. Next, include the check boxes:

```
<CheckBox
    android:id="@+id/check_egg"
    android:layout_width="wrap_content"
    android:layout_height="wrap_content"
    android:checked="false"
    android:paddingBottom="@dimen/padding"
    android:text="@string/egg" />

<CheckBox
    android:id="@+id/check_cress"
    android:layout_width="wrap_content"
    android:layout_height="wrap_content"
    android:checked="false"
    android:paddingBottom="@dimen/padding"
    android:text="@string/cress" />
```

4. Then add the switch:

```
<Switch
    android:id="@+id/switch_salt"
    android:layout_width="wrap_content"
    android:layout_height="wrap_content"
    android:checked="false"
    android:paddingBottom="@dimen/padding"
    android:paddingTop="@dimen/padding"
    android:text="@string/salt" />
```

5. This is an inner relative layout containing the following action buttons:

```
<TextView
    android:id="@+id/action_ok"
    android:layout_width="wrap_content"
    android:layout_height="wrap_content"
    android:layout_alignParentEnd="true"
    android:layout_gravity="end"
    android:background="?attr/selectableItemBackground"
    android:clickable="true"
    android:gravity="center_horizontal"
    android:minWidth="@dimen/action_minWidth"
    android:onClick="onActionOkClicked"
    android:padding="@dimen/padding"
    android:text="@android:string/ok"
    android:textColor="@color/colorAccent" />

<TextView
    android:id="@+id/action_cancel"
    android:layout_width="wrap_content"
    android:layout_height="wrap_content"
    android:layout_gravity="end"
    android:layout_toStartOf="@id/action_ok"
    android:background="?attr/selectableItemBackground"
    android:clickable="true"
    android:gravity="center_horizontal"
    android:minWidth="@dimen/action_minWidth"
    android:padding="@dimen/padding"
    android:text="@string/action_cancel_text"
    android:textColor="@color/colorAccent" />
```

Note the use of `android:onClick="onActionOkClicked"` in the OK button. This can be used in lieu of a click listener and identifies the method on the owning activity to be called when the view is clicked on. This is a very convenient technique, although it does rather blur the lines between model and view and can be prone to bugs.

Before we add this method, we need to declare and instantiate one or two fields and views. Follow these steps to complete the exercise:

1. Include the following field declarations in the class:

```
public SandwichBuilder builder;
public Sandwich sandwich;

private  RadioButton bagel;
public CheckBox egg, cress;
public Switch salt;
public TextView order;
```

2. Instantiate the widgets like so:

```
bagel = (RadioButton) findViewById(R.id.radio_bagel);
egg = (CheckBox) findViewById(R.id.check_egg);
cress = (CheckBox) findViewById(R.id.check_cress);
salt = (Switch) findViewById(R.id.switch_salt);
order = (TextView) findViewById(R.id.text_order);
```

3. Now we can add the onActionOkClicked() method we declared in the XML layout:

```
public void onActionOkClicked(View view) {
    builder = new SandwichBuilder();
    sandwich = new Sandwich();

    // Radio button group
    if (bagel.isChecked()) {
        sandwich = builder.build(sandwich, new Bagel());
    } else {
        sandwich = builder.build(sandwich, new Bun());
    }

    // Check boxes
    if (egg.isChecked()) {
        sandwich = builder.build(sandwich, new Egg());
    }

    if (cress.isChecked()) {
        sandwich = builder.build(sandwich, new Cress());
    }

    // Switch
    if (salt.isChecked()) {
        sandwich = builder.build(sandwich, new Salt());
```

```
    }

    // Display output
    order.setText(new StringBuilder()
            .append(sandwich.getSandwich())
            .append("\n")
            .append(sandwich.getKcal())
            .append(" kcal")
            .toString());
}
```

We can now test this code on a device, and despite the small number of ingredients, it should be clear how this works to allow users to build sandwiches of their choice:

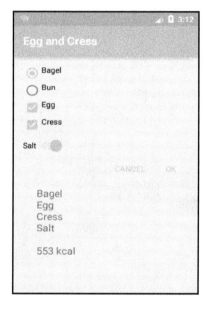

Multiple widgets

We only need to include more ingredients and a more sophisticated UI to handle this. Nevertheless, the principle will remain the same and the same structure and logic can be applied.

Despite the potential, the preceding example lacks the decorative features we saw earlier, such as offering toasted varieties and low-fat spread. Fortunately, it is a simple task to attach decorators to both our bread and filling classes. Before we do so, we will take a quick look at why a builder is not the only candidate pattern capable of performing this task.

Selecting patterns

Examine the following figure comparing a builder and an abstract factory:

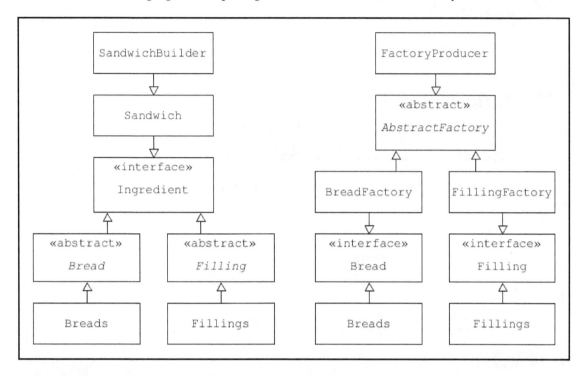

Comparison between the builder and abstract factory patterns

Despite the differences in approach, there are striking similarities between the builder and abstract factory patterns, and they both perform similar functions. We could quite easily use abstract factories for this task. Factories are more flexible when it comes to adding or modifying products, and are structurally a little simpler, but there is one important difference between the two patterns that really determines our choice.

Both factories and builders manufacture objects, but the major difference is that factories return their products as each of them are requested. This would be like having a sandwich delivered one ingredient at a time. The builder, on the other hand, only constructs its output once all the products have been selected, and this is far more like the behavior of making and delivering a sandwich. This is the reason why the builder pattern provides the best solution in this case. With this decision taken, we can stick with the preceding code and get down to adding a little extra functionality.

Adding a decorator

As we know, one of the best ways to add further functionality is with decorator patterns. We have already seen how these work, and now we can add one to our simple sandwich builder. Individual decorations are almost identical in structure, differing only in the values they return, so we need to create only one here, by way of example.

Attaching the pattern

Follow these steps to add the option to offer a toasted sandwich:

1. Open the empty `Bread` class and complete it like so:

```
public abstract class Bread implements Ingredient {

    String decoration;
    int decorationKcal;

    public String getDecoration() {
        return decoration;
    }

    public int getDecorationKcal() {
        return decorationKcal;
    }
}
```

2. Create a `BreadDecorator` class like the one found here:

```
public abstract class BreadDecorator extends Bread {

    public abstract String getDecoration();

    public abstract int getDecorationKcal();
}
```

3. Now add the concrete decorator itself:

```
public class Toasted extends BreadDecorator {
    private Bread bread;

    public Toasted(Bread bread) {

        this.bread = bread;
    }
```

```
@Override
public String getDecoration() {

    return "Toasted";
}

@Override
public int getDecorationKcal() {

    return 0;
}

// Required but not used
@Override
public String description() { return null; }

@Override
public int kcal() { return 0; }
}
```

Not only does the use of a decorator keep the number of sub-classes we need to a minimum, it also serves a perhaps more useful function, in that it allows us to include options such as toasted and/or open, which are not ingredients, strictly speaking, and this helps keep our classes meaningful.

It should be clear that we can now add as many such decorations as we like, but first there are, of course, one or two changes we need to make to our main source code to see our decoration in action.

Connecting the pattern to the UI

Edit the main XML layout and Java activity to achieve this by following these simple steps:

1. Add the following switch, just below the radio button group:

```
<Switch
    android:id="@+id/switch_toasted"
    android:layout_width="wrap_content"
    android:layout_height="wrap_content"
    android:checked="false"
    android:paddingBottom="@dimen/padding"
    android:paddingTop="@dimen/padding"
    android:text="@string/toasted" />
```

2. Open the `MainActivity` class and provide it these two fields:

```
public Switch toasted;
public Bread bread;
```

3. Instantiate the widget like so:

```
toasted = (Switch) findViewById(R.id.switch_toasted);
```

4. Add the following method variables to the `onActionOkClicked()` method:

```
String toast;
int extraKcal = 0;
```

5. Now add this code underneath the radio buttons:

```
// Switch : Toasted
if (toasted.isChecked()) {
    Toasted t = new Toasted(bread);
    toast = t.getDecoration();
    extraKcal += t.getDecorationKcal();
} else {
    toast = "";
}
```

6. Finally, modify the text output code like so:

```
order.setText(new StringBuilder()
        .append(toast + " ")
        .append(sandwich.getSandwich())
        .append("\n")
        .append(sandwich.getKcal() + extraKcal)
        .append(" kcal")
        .append("\n")
        .toString());
```

That is all that is needed to add a decorator to our existing pattern and provide it as a working part of our UI.

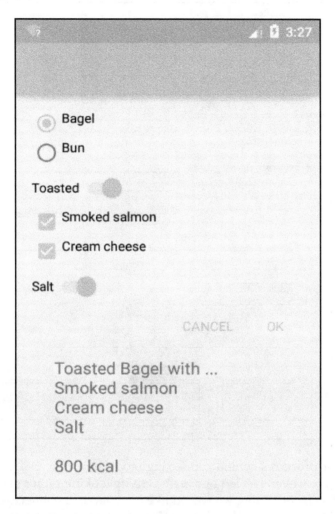

Note that although the filling classes have been refactored here to something tastier, the code, however, remains the same. Everything from variables to classes and packages can be refactored with Shift + F6. This will also rename all occurrences, calls, and even getters and setters.
To rename an entire project, rename the directory in your Android Studio projects folder and then open it from the File menu.

As a UML class diagram, we can express this new structure like so:

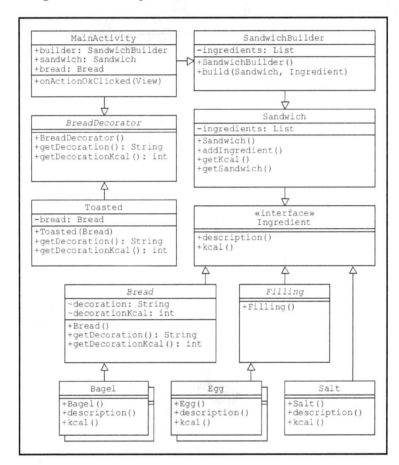

This covers the basic processes behind connecting model and view using simple design patterns. Our work, however, has left our main activity looking rather messy and complicated, and this is something we would like to avoid. It is not necessary to implement this here, as this is still a very simple program. However, there will be times when client code can become very cluttered with listeners and various other callbacks, and it is useful to know how best to square things away using a pattern.

The facade is the most useful pattern for this kind of thing, and is quick and easy to implement. We have come across this pattern before, and implementing it here is left as an exercise for the reader. The class structure would look something like this:

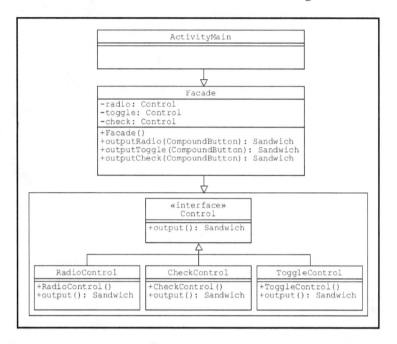

Summary

In this chapter, we have seen how to combine design patterns to perform complex tasks. We created a builder to allow users to construct a sandwich of their choice and to customize it with a decorator pattern. We also explored another vital pattern, the prototype, and saw how vital it can be whenever we have large files or slow processes to contend with.

As well as delving into the concepts of pattern design, the chapter included the more practical aspects of setting, reading, and responding to compound buttons such as switches and checkboxes, and this formed a significant step on the way to developing more sophisticated systems.

In the next chapter, we will look more closely at communicating with the user through the use of various Android notification tools, such as the snack bar and how services and broadcasts play a part in Android development.

8
Composing Patterns

We have seen how patterns can be used to manipulate, organize, and present data, but this data has been fleeting, and we have not yet considered how to ensure data persists from one session to the next. In this chapter, we will look at how this is done using internal data storage mechanisms. In particular, we will explore how users can save their preferences, making the app simpler and more fun to use. Before we do this, we will begin the chapter by examining the composite pattern and its uses, particularly when it comes to constructing hierarchical structures such as Android UIs.

In this chapter, you will learn how to do the following:

- Construct a composite pattern
- Create a layout with composer
- Use static files
- Edit application files
- Store user preferences
- Understand the activity life cycle
- Add unique identifiers

One of the most immediate ways we can apply design patterns to an Android project is layout inflation, and back in Chapter 6, *Activating Patterns*, we used a builder pattern to inflate a simple layout. This example had some serious shortcomings. It only handled text views and did not cater for nested layouts. For dynamic layout inflation to be of real use to us, we need to be able to include any type of widget or view at any level of the layout hierarchy, and this is where the composite design pattern comes in.

The composite pattern

At first glance, the composite pattern might seem very similar to the builder pattern, as both build complex objects out of smaller ones. There is, however, a significant difference in the approach these patterns take. Builders work in a very linear fashion, adding objects one at a time. The composite pattern, on the other hand, can add groups of objects as well as individual ones. More importantly, it does so in such a way that the client can add individual objects or groups of objects without having to concern itself with which it is dealing with. In other words, we can add completed layouts, individual views, or groups of views with exactly the same code.

Along with the ability to compose branching data structures, the ability to hide from the client the details of the objects being manipulated is what makes the composer pattern so powerful.

Before creating a layout composer, we will take a look at the pattern itself, applied to a very simple model, so that we can appreciate the workings of the pattern better. Here is the overall structure. As you can see, it is very simple conceptually.

Follow these steps to construct our composite pattern:

1. Start with an interface that can represent both individual components and collections of components, like so:

```
public interface Component {

    void add(Component component);
    String getName();
    void inflate();
}
```

2. Add this class to extend the interface for individual components:

```
public class Leaf implements Component {
    private static final String DEBUG_TAG = "tag";
    private String name;

    public Leaf(String name) {
        this.name = name;
    }

    @Override
    public void add(Component component) { }

    @Override
```

```java
    public String getName() {
        return name;
    }

    @Override
    public void inflate() {
        Log.d(DEBUG_TAG, getName());
    }
}
```

3. Then add the class for collections:

```java
public class Composite implements Component {
    private static final String DEBUG_TAG = "tag";

    // Store components
    List<Component> components = new ArrayList<>();
    private String name;

    public Composite(String name) {
        this.name = name;
    }

    @Override
    public void add(Component component) {
        components.add(component);
    }

    @Override
    public String getName() {
        return name;
    }

    @Override
    public void inflate() {
        Log.d(DEBUG_TAG, getName());

        // Inflate composites including children
        for (Component component : components) {
            component.inflate();
        }
    }
}
```

As you can see here, this pattern is very simple, yet very effective:

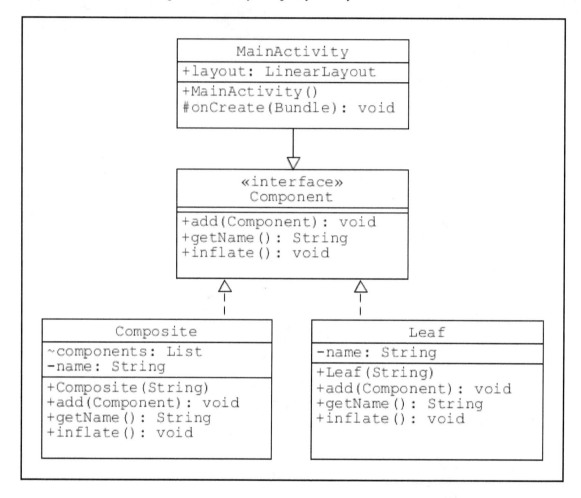

To see it in action, we need to define some components and composites. We can define components with lines like this:

```
Component newLeaf = new Leaf("New leaf");
```

We can create composite collections with the `add()` method like so:

```
Component composite1 = new Composite("New composite");
composite1.add(newLeaf);
composite1.add(oldLeaf);
```

Nesting compositions within each other is just as simple because we have written the code so that we can disregard whether we are creating a `Leaf` or a `Composite` and use the same code for either. Here's an example:

```
Component composite2 = Composite("Another composite");
composite2.add(someLeaf);
composite2.add(composite1);
composite2.add(anotherComponent);
```

Displaying a component, which is simply text in this case, is as easy as calling its `inflate()` method.

Adding a builder

Defining and printing a fair selection of outputs would make for considerably cluttered client code, and the solution we will adopt here is to steal an idea from another pattern and employ a single builder class to do the work of constructing our desired compositions. These could be anything we like, and here is one possible builder:

```
public class Builder {

    // Define individual components
    Component image = new Leaf("  image view");
    Component text = new Leaf("  text view");
    Component list = new Leaf("  list view");

    // Define composites
    Component layout1(){
        Component c = new Composite("layout 1");
        c.add(image);
        c.add(text);
        return c;
    }

    // Define nested composites
    Component layout2() {
        Component c = new Composite("layout 2");
        c.add(list);
        c.add(layout1());
        return c;
    }

    Component layout3(){
        Component c = new Composite("layout 3");
        c.add(layout1());
```

```
        c.add(layout2());
        return c;
    }
}
```

This leaves the `onCreate()` method of our activity clean and simple to follow, as you can see here:

```
@Override
protected void onCreate(Bundle savedInstanceState) {

    super.onCreate(savedInstanceState);
    setContentView(R.layout.activity_main);

    Builder builder = new Builder();

    // Inflate a single component
    builder.list.inflate();

    // Inflate a composite component
    builder.layout1().inflate();

    // Inflate nested components
    builder.layout2().inflate();
    builder.layout3().inflate();
}
```

Although we have only produced a basic output, it should be clear how we can now extend this to inflating real layouts and how useful this technique can be.

A Layout composer

In Chapter 6, *Activating Patterns*, we used a builder to construct a simple UI. The builder was a perfect choice for this task, as we were only concerned with including one type of view. We could have adapted this scheme (literally, with an adapter) to cater for other view types, but it would be far better to employ a pattern that does not care what type of component it is dealing with. Hopefully, the preceding example demonstrates the composite pattern's suitability to this kind of task.

In the following example, we will apply the same principle to an actual UI inflater that works with different types of view, composite groups of views and, most significantly, dynamic nested layouts.

For the purpose of this exercise, we will suppose that our app has a news page. This would largely be a promotional feature, but it has been demonstrated that consumers are more susceptible to advertising when it is dressed up as news. Many of the components, such as header and logo, will remain static, while others will change frequently in both content and layout structure. This makes it an ideal subject for our composer pattern.

Here is the UI we will be developing:

Adding components

We will approach each problem individually, building the code as we go. First, we will tackle the issue of creating and displaying a single component view by following these steps:

1. As before, we start with the `Component` interface:

```
public interface Component {

    void add(Component component);
    void setContent(int id);
    void inflate(ViewGroup layout);
}
```

2. Now implement this with the following class:

```
public class TextLeaf implements Component {
    public TextView textView;

    public TextLeaf(TextView textView, int id) {
        this.textView = textView;
        setContent(id);
    }

    @Override
    public void add(Component component) { }

    @Override
    public void setContent(int id) {
        textView.setText(id);
    }

    @Override
    public void inflate(ViewGroup layout) {
        layout.addView(textView);
    }
}
```

3. Next, add the `Builder`, which, for now, is remarkably simple, containing only two properties and a constructor:

```
public class Builder {
    Context context;
    Component text;

    Builder(Context context) {
```

```
        this.context = context;
        init();
        text = new TextLeaf(new TextView(context),
                R.string.headline);
    }
}
```

4. Finally, edit the `onCreate()` method of the activity to use our own layout as the root and add our view to it, like this:

```
@Override
protected void onCreate(Bundle savedInstanceState) {

    super.onCreate(savedInstanceState);

    // Replace default layout
    LinearLayout layout = new LinearLayout(this);

    layout.setOrientation(LinearLayout.VERTICAL);
    layout.setLayoutParams(new ViewGroup.LayoutParams(
            ViewGroup.LayoutParams.MATCH_PARENT,
            ViewGroup.LayoutParams.WRAP_CONTENT));
    setContentView(layout);

    // Add component
    Builder builder = new Builder(this);
    builder.headline.inflate(layout);
}
```

As it stands, there is nothing impressive about what we have done here, but having worked through the previous example, it will be clear where we are going with this, and our next step is to create a component that will handle image views.

As you can see in the following snippet, the `ImageLeaf` class is almost identical to its textual sibling, differing only in the view type it generates and the use of `setImageResource()` to operate on the `id` argument:

```
public class ImageLeaf implements Component {
    private ImageView imageView;

    public ImageLeaf(ImageView imageView, int id) {
        this.imageView = imageView;
        setContent(id);
    }

    @Override
    public void add(Component component) { }
```

```
        @Override
        public void setContent(int id) {
            imageView.setImageResource(id);
        }

        @Override
        public void inflate(ViewGroup layout) {
            layout.addView(imageView);
        }
    }
```

This can be easily added to the builder in the same way as the text view, although now we will create a small method for this and call it from the constructor, as we may want to add a number of others. The code should now look like this:

```
Builder(Context context) {
    this.context = context;
    initLeaves();
}

private void initLeaves() {

    header = new ImageLeaf(new ImageView(context),
            R.drawable.header);

    headline = new TextLeaf(new TextView(context),
            R.string.headline);
}
```

As intended, as far as the client code is concerned, there is no difference between this and any other component, and it can be inflated with this:

```
builder.header.inflate(layout);
```

Image views and text views can both take their respective primary contents (image and text) as resource ID integers and so we can use the same `int` parameter for both. Dealing with this in the `setContent()` method allows us to decouple the actual implementations and allow us to reference each of them simply as a `Component`. The `setContent()` method will also prove useful shortly when we apply some formatting attributes.

This is all still very basic, and if we were to create all our components like this, regardless of how they were grouped into compositions, the builder code would soon become very long-winded. The banner views we just created are unlikely to change, and so this system suits this setup. However, we will need to find a more flexible approach for more variable content, but before we do, we will see how to create composite versions of our classes.

Creating composites

The true usefulness of the composite pattern lies in its ability to treat groups of objects as one, and our two header views offer a good opportunity to show how. Seeing as they always appear together, it makes perfect sense to treat them as one. There are three ways we can do this:

- Adapt one of the existing leaf classes so it can create children
- Create a composite with no parent
- Create a composite with a layout as parent

We will see how to do all of these, but first we will implement the most efficient in this case and create a composite class based on one of our leaf classes. We want the image of our header above the text so will will use the `ImageLeaf` class as our template.

There are just three quick steps to this task:

1. The `CompositeImage` class is identical to `ImageLeaf`, with the following exceptions:

```
public class CompositeImage implements Component {
    List<Component> components = new ArrayList<>();

    ...

    @Override
    public void add(Component component) {
        components.add(component);
    }

    ...

    @Override
    public void inflate(ViewGroup layout) {
        layout.addView(imageView);

        for (Component component : components) {
            component.inflate(layout);
        }
    }
}
```

2. Building this group in the builder is as simple as this:

```
Component headerGroup() {
    Component c = new CompositeImage(new ImageView(context),
            R.drawable.header);
    c.add(headline);
    return c;
}
```

3. We can now replace our calls in the activity too:

```
builder.headerGroup().inflate(layout);
```

This too can be treated exactly the same as all the other components, and making an equivalent text version would be very simple. Classes such as these can be seen as extensions of their leaf versions, and it was useful here, but it would be tidier to create a composite with no container, as this would enable us to organize groups that we can later insert into a layout.

The following class is a pared-down composite class that can be used to group any components, including other groups:

```
class CompositeShell implements Component {
    List<Component> components = new ArrayList<>();

    @Override
    public void add(Component component) {
        components.add(component);
    }

    @Override
    public void setContent(int id) { }

    @Override
    public void inflate(ViewGroup layout) {

        for (Component component : components) {
            component.inflate(layout);
        }
    }
}
```

Imagine we want to group three images to be added later to a layout. The way the code stands, we would have to add these definitions during construction. This could result in some bulky and unattractive code. We will solve this here by simply adding methods to the builder that allow us to create components as we need them.

These two methods are as follows:

```
public TextLeaf setText(int t) {
    TextLeaf leaf = new TextLeaf(new TextView(context), t);
    return leaf;
}


public ImageLeaf setImage(int t) {
    ImageLeaf leaf = new ImageLeaf(new ImageView(context), t);
    return leaf;
}
```

We can use the builder to construct these groups like so:

```
Component sandwichArray() {
    Component c = new CompositeShell();

    c.add(setImage(R.drawable.sandwich1));
    c.add(setImage(R.drawable.sandwich2));
    c.add(setImage(R.drawable.sandwich3));
    return c;
```

This group can be inflated like any other component from the client and, because our layout has vertical orientation, will be displayed as a column. If we want them output in a row, we will need a horizontal orientation and therefore a class to generate one.

Create composite layouts

Here is the code for a composite component that will generate a linear layout as its root and place any added views inside itself:

```
class CompositeLayer implements Component {
    List<Component> components = new ArrayList<>();
    private LinearLayout linearLayout;

    CompositeLayer(LinearLayout linearLayout, int id) {
        this.linearLayout = linearLayout;
        setContent(id);
    }

    @Override
    public void add(Component component) {
        components.add(component);
    }
```

```
    @Override
    public void setContent(int id) {
        linearLayout.setBackgroundResource(id);
        linearLayout.setOrientation(LinearLayout.HORIZONTAL);
    }

    @Override
    public void inflate(ViewGroup layout) {
        layout.addView(linearLayout);

        for (Component component : components) {
            component.inflate(linearLayout);
        }
    }
}
```

The code to construct this in the builder is no different to the others:

```
Component sandwichLayout() {
    Component c = new CompositeLayer(new LinearLayout(context),
            R.color.colorAccent);
    c.add(sandwichArray());
    return c;
}
```

We can now inflate our compositions with just a couple of clearly understandable lines of code in our activity:

```
Builder builder = new Builder(this);
builder.headerGroup().inflate(layout);
builder.sandwichLayout().inflate(layout);
```

It is worth noting how we used the setContent() method of the composite layer to set the orientation. Looking at the overall structure, this is clearly the right place to do this and this brings us on to our next task, formatting our UI.

Formatting layouts at runtime

Although we now have the means to produce any number of complex layouts, a quick look at the following output demonstrates that in terms of appearance and design, we are still a long way from our desired design:

We saw previously how we set an inserted layout's orientation from its `setContent()` method and this is how we can take greater control over our components' appearance. Taking this further it only takes a moment or two to produce an acceptable layout. Just follow these simple steps:

1. Begin by editing the `setContent()` method of the `TextLeaf`, like this:

```
@Override
public void setContent(int id) {

    textView.setText(id);

    textView.setPadding(dp(24), dp(0), dp(0), dp(16));
    textView.setTextSize(TypedValue.COMPLEX_UNIT_SP, 24);
    textView.setLayoutParams(new ViewGroup.LayoutParams(
            ViewGroup.LayoutParams.MATCH_PARENT,
            ViewGroup.LayoutParams.WRAP_CONTENT));
}
```

2. This will also require the following method for converting from px to dp:

```
private int dp(int px) {
    float scale = textView.getResources()
            .getDisplayMetrics()
            .density;
    return (int) (px * scale + 0.5f);
}
```

3. The `ImageLeaf` component just requires these changes:

```
@Override
public void setContent(int id) {
    imageView.setScaleType(ImageView.ScaleType.FIT_CENTER);

    imageView.setLayoutParams(new ViewGroup.LayoutParams(
            ViewGroup.LayoutParams.WRAP_CONTENT,
            dp(R.dimen.imageHeight)));

    imageView.setImageResource(id);
}
```

4. We also added one more construction to the builder, like so:

```
Component story(){
    Component c = new CompositeText(new TextView(context)
            ,R.string.story);
    c.add(setImage(R.drawable.footer));
```

```
        return c;
}
```

5. This can now be put into place with the follow lines in the activity:

```
Builder builder = new Builder(this);

builder.headerGroup().inflate(layout);
builder.sandwichLayout().inflate(layout);
builder.story().inflate(layout);
```

These adjustments should now produce a design along the lines of our original spec. Although we have added a lot of code and created specific Android objects, a look at the following diagram will demonstrate that the overall pattern remains the same:

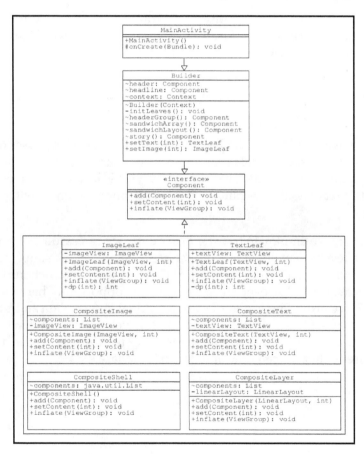

There is a great deal more that we could do here, for example, tackling the issue of developing landscape layouts and scaling for different screen configurations, all of which could be managed simply using the same methodology. What we have done, however, is enough to demonstrate how to construct a layout dynamically at runtime using a composite pattern.

We are going to leave this pattern for now, as we explore how to provide a little customization functionality and consider user preferences and how we can store persistent data.

Storage options

Nearly all apps have some form off **settings** menu that allow users to store regularly accessed information and customize the app to their own preferences. These settings can be anything from changing a password to personalizing the color scheme or any number of other tweaks and adjustments.

If you have a lot of data and access to a web server, it is often best to cache data from this source, as this will both save battery and speed up the application.

First and foremost, we should consider how these setting can save our user time. No one wants to have to enter all their details every time they order a sandwich, nor do they want to have to construct the same sandwich over and over. This leads to the question of how we represent a sandwich across the system and how order information is sent to and received by the vendor.

Whatever technology we employ to transfer order data, we can assume at some point in the process there will be a human being making the actual sandwich. A simple text string would seem to be all we need, and it would certainly suffice as instructions for the vendor and to store user favorites. However, there is a valuable opportunity here that it would be foolish to miss. Every order that is placed contains valuable sales data and by collating these we can build up a picture of which products sell well and which don't. For this reason, we need to include as much data as we can in the order message. Purchase history can contain a lot of useful data, as can time and date of purchase.

Whatever supporting data we choose to collect, one thing that would be immensely useful would be to able to identify individual customers, but people do not like giving out personal information, and nor should they. There is no reason why someone would need to give out their date of birth or gender, simply to buy a sandwich. However, as we shall see, we can attach a unique identifier to each downloaded app and/or the device it is running on. Furthermore, there is no way that we, or anyone else, can identify individuals from this, and it is therefore no threat to their security or privacy, which it is vital that we protect.

There are several ways we can store data on a user's device so that properties persist across sessions. Usually, we will want this data kept private, and in the next section we will see how this can be achieved.

Creating static files

Our main focus in this part of the chapter will be the storage of user preferences. There are one or two other storage options we should take a look beforehand, starting with the device's internal storage.

In the first half of the chapter, we assigned quite a long string using the `strings.xml` values file. This, and similar, resource files are best suited to storing individual words and short phrases, but form an unattractive method for storing long sentences or paragraphs. For these circumstances, we can use text files and store them in the `res/raw` directory.

The handy thing about the `raw` directory is that it is compiled as part of the R class, which means its contents can be referenced in the same way as any other resource, such as a string or drawable, for example, `R.raw.some_text`.

To see how to include lengthy text without messing up the strings file, follow these simple steps:

1. The `res/raw` folder is not included by default, so begin by creating it.
2. Create a new file in this folder containing your text. Here, it is called `wiki`, as it is taken from the Wikipedia entry for sandwich.
3. Open your activity or whichever code you are using to inflate your layout and add this method:

```java
public static String readFile(Context context, int resId) {

    InputStream stream = context.getResources()
            .openRawResource(R.raw.wiki);
    InputStreamReader inputReader = new InputStreamReader(stream);
    BufferedReader bufferedReader = new BufferedReader(inputReader);
    String line;
    StringBuilder builder = new StringBuilder();

    try {
        while ((line = bufferedReader.readLine()) != null) {
            builder.append(line)
                    .append('\n');
        }
    } catch (IOException e) {

        return null;
    }

    return builder.toString();
}
```

4. Now simply add these lines to populate your view.

```java
TextView textView = (TextView) findViewById(R.id.text_view);
String data = readFile(this, R.raw.wiki);
textView.setText(data);
```

One of the nice things about the raw folder being treated like other resource directories is that we can create designated versions for different devices or locales. For example, here we have created a folder called `raw-es` and placed a Spanish translation of the text inside with the same name:

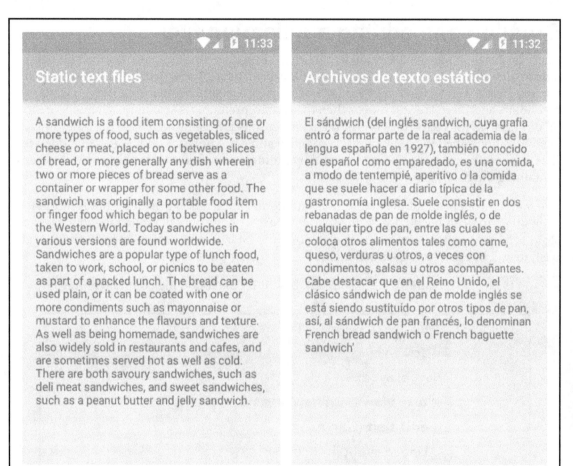

If you are using an external text editor, such as Notepad, you will need to ensure that the file is saved in UTF-8 format for the non-Latin characters to display correctly.

This kind of resource is very useful and very easy to implement, but such files are read-only, and there are bound to be times when we would like to create and edit this kind of file.

Creating and editing application files

There is of course far more we can do here than store long strings conveniently, and being able to alter the content of such files at runtime gives us a lot of scope. If there weren't already such a a convenient method for storing user preferences, it would make a good candidate, and there are still times when the shared preferences structure is inadequate for all our needs. This constitutes one of the main reasons for using such files; the other is as a customization function, allowing users to make and store notes or bookmarks. Encoded text files can even be created to be understood by builders and used to rebuild sandwich objects containing the user's favorite ingredients.

The method we are about to explore uses an internal application directory that is hidden from other apps on the device. In the following exercise, we will demonstrate how users can store persistent and private text files using our app. Start a new project or open one that you wish to add internal storage functionality to and follow these steps:

1. Start by creating a simple layout. Base it on the following component tree:

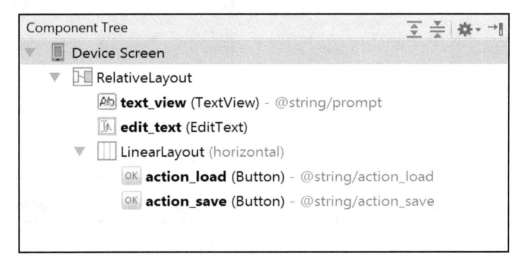

2. For simplicity's sake, we will use the XML onClick property to assign code to each button using `android:onClick="loadFile"` and `android:onClick="saveFile"` respectively.

3. First, construct the `saveFile()` method:

```
public void saveFile(View view) {

    try {
        OutputStreamWriter writer = new
OutputStreamWriter(openFileOutput(fspc, 0));
        writer.write(editText.getText().toString());
        writer.close();

    } catch (IOException e) {
        e.printStackTrace();
    }
}
```

4. Then make the `loadFile()` method:

```
public void loadFile(View view) {

    try {
        InputStream stream = openFileInput(fspc);
        if (stream != null) {
            InputStreamReader inputReader = new InputStreamReader(stream);
            BufferedReader bufferedReader = new
BufferedReader(inputReader);
            String line;
            StringBuilder builder = new StringBuilder();

            while ((line = bufferedReader.readLine()) != null) {
                builder.append(line)
                        .append("\n");
            }

            stream.close();
            editText.setText(builder.toString());
        }

    } catch (IOException e) {
        e.printStackTrace();
    }
}
```

This example is very simple, but it only needs to demonstrate the potential of being able to store data this way. Using the preceding layout, the code is easy to test.

Storing user data, or data we want on the user, like this is very handy and very safe. We could always encrypt this data too, but that is a subject for another book. The Android framework is no more or less secure than any other mobile platform, and as we will not be storing anything more sensitive than preferences in sandwich fillings, this system will suit our purposes just fine.

It is, of course, also possible to create and access files on a device's external storage, such as a micro SD card. These files are public by default and are usually created when we want to share something with other apps. The process is similar to those we have just explored, and so we won't cover it here. Instead, we will get on with storing user preferences using the built in **SharedPreferences** interface.

Storing user preferences

We have already covered why being able to store user settings is so important, and we thought briefly about what settings we would like to store. Shared preferences use key-value pairs to store their data, and this is fine for values such as `name="desk" value="4"` but we want some quite detailed information about some things. For example, we want the user to be able to store their favorite sandwiches for easy recall.

The first step with this is to see how the Android shared preferences interface works generally and where it should be applied.

The activity life cycle

Storing and retrieving user preferences using the **SharedPreferences** interface uses key-value pairs to store and retrieve primitive data types. This is very simple to apply, and the process only really gets interesting when we ask when and where we should perform these actions. This brings us to activity life cycles.

Unlike desktop applications, mobile apps are not usually closed down deliberately by the user. Instead, they are generally navigated away from, often leaving them semi-active in the background. During runtime, an activity will enter a variety of states, such as paused, stopped, or resumed. Each of these states has an associated callback method, such as the `onCreate()` method that we are more than familiar with. There are several of these that we could use to save and load our user settings, and to decide which, we need to take a look at the life cycle itself:

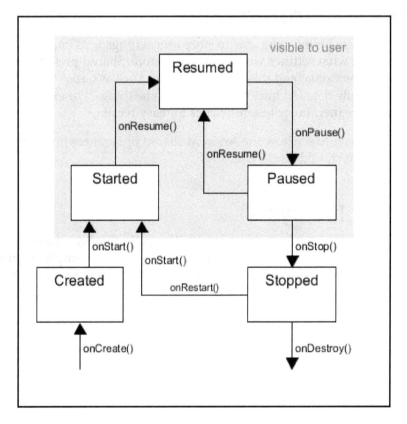

The preceding diagram can appear a little confusing, and the best way to see what happens when is to write a little debug code. Including `onCreate()`, there are seven callback methods that can be called during an activity's life time:

- onCreate()
- onStart()
- onResume()
- onPause()

- onStop()
- onDestroy()
- onRestart()

At first, it might seem to make sense to save user settings from the `onDestroy()` method, as it is the last possible state. To see why this does not often work, open any project and override each of the methods in the preceding list and add some debug code, as seen in the example here:

```
@Override
public void onResume() {
    super.onResume();
    Log.d(DEBUG_TAG, "Resuming...");
}
```

A few moments experimentation is enough to see that `onDestroy()` is not always called. To ensure our data get saved, we need to store our preferences from the `onPause()` or `onStop()` methods.

Applying preferences

To see how preferences are stored and retrieved, start a new project or open an existing one and follow these steps:

1. First, create a new class, `User`, like so:

```
// Singleton class as only one user
public class User {
    private static String building;
    private static String floor;
    private static String desk;
    private static String phone;
    private static String email;
    private static User user = new User();

    public static User getInstance() {
        return user;
    }

    public String getBuilding() {
        return building;
    }

    public void setBuilding(String building) {
```

```
            User.building = building;
        }

        public String getFloor() {
            return floor;
        }

        public void setFloor(String floor) {
            User.floor = floor;
        }

        public String getDesk() {
            return desk;
        }

        public void setDesk(String desk) {
            User.desk = desk;
        }

        public String getPhone() {
            return phone;
        }

        public void setPhone(String phone) {
            User.phone = phone;
        }

        public String getEmail() {
            return email;
        }

        public void setEmail(String email) {
            User.email = email;
        }
    }
```

2. Next, create an XML layout to match this data based on the following preview:

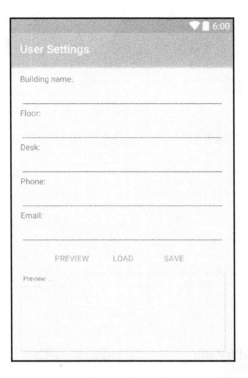

3. Modify the activity so that it implements the following listener.

```
public class MainActivity
    extends AppCompatActivity
    implements View.OnClickListener
```

4. Include the following fields and associate them with their XML counterparts in the usual manner:

```
private User user = User.getInstance();

private EditText editBuilding;
private EditText editFloor;
private EditText editDesk;
private EditText editPhone;
private EditText editEmail;

private TextView textPreview;
```

5. Add the buttons locally in the `onCreate()` method and set their click listeners:

```
Button actionLoad = (Button) findViewById(R.id.action_load);
Button actionSave = (Button) findViewById(R.id.action_save);
Button actionPreview = (Button) findViewById(R.id.action_preview);

actionLoad.setOnClickListener(this);
actionSave.setOnClickListener(this);
actionPreview.setOnClickListener(this);
```

6. Create the following method and call it from within `onCreate()`:

```
public void loadPrefs() {
    SharedPreferences prefs = getApplicationContext()
        .getSharedPreferences("prefs", MODE_PRIVATE);

    // Retrieve settings
    // Use second parameter if never saved
    user.setBuilding(prefs.getString("building", "unknown"));
    user.setFloor(prefs.getString("floor", "unknown"));
    user.setDesk(prefs.getString("desk", "unknown"));
    user.setPhone(prefs.getString("phone", "unknown"));
    user.setEmail(prefs.getString("email", "unknown"));
}
```

7. Create a method for storing the preferences like this:

```
public void savePrefs() {
    SharedPreferences prefs =
getApplicationContext().getSharedPreferences("prefs", MODE_PRIVATE);
    SharedPreferences.Editor editor = prefs.edit();

    // Store preferences
    editor.putString("building", user.getBuilding());
    editor.putString("floor", user.getFloor());
    editor.putString("desk", user.getDesk());
    editor.putString("phone", user.getPhone());
    editor.putString("email", user.getEmail());

    // Use apply() not commit()
    // to perform operation in background
    editor.apply();
}
```

8. Add the `onPause()` method to call it:

```
@Override
public void onPause() {
    super.onPause();
    savePrefs();
}
```

9. Finally add the click listener, like so:

```
@Override
public void onClick(View view) {

    switch (view.getId()) {

        case R.id.action_load:
            loadPrefs();
            break;

        case R.id.action_save:
            // Recover data from form
            user.setBuilding(editBuilding.getText().toString());
            user.setFloor(editFloor.getText().toString());
            user.setDesk(editDesk.getText().toString());
            user.setPhone(editPhone.getText().toString());
            user.setEmail(editEmail.getText().toString());
            savePrefs();
            break;

        default:
            // Display as string
            textPreview.setText(new StringBuilder()
                    .append(user.getBuilding()).append(", ")
                    .append(user.getFloor()).append(", ")
                    .append(user.getDesk()).append(", ")
                    .append(user.getPhone()).append(", ")
                    .append(user.getEmail()).toString());
            break;
    }
}
```

The load and preview functions have been added here simply to allow us to test our code, but as you can see, this process can be used to store and retrieve any amount of pertinent data:

 If for any reason you need to empty your preferences file, this can be done with the `edit.clear()` method.

It is quite possible to locate and look at our shared preferences thanks to the Android Device Monitor, which can be accessed through the **Tools | Android** menu. Open the **File explorer** and navigate to `data/data/com.your_app/shared_prefs/prefs.xml`. It should look something like this:

```
<?xml version='1.0' encoding='utf-8' standalone='yes' ?>
<map>
    <string name="phone">+44 0102 555 6789</string>
    <string name="email">kyle@blt.com</string>      <string
```

```
name="floor">5</string>
    <string name="desk">13</string>       <string name="user_id">
        fbc08fca-f375-4786-9e2d-d610c9cd0377</string>
    <boolean name="new_user" value="false" />       <string
name="building">Bagel Building</string> </map>
```

Despite its simplicity, shared preferences form an essential element of nearly all Android mobile apps, and as well as these obvious advantages, there is one other neat trick we can perform here. We can use the content of the shared preferences file to determine whether an app is being run for the first time.

Adding a unique identifier

It is always a good idea when gathering sales data to have some means of identifying individual customers. This need not be by name or anything personal, a simple ID number can add a whole new dimension to a data set.

In many situations, we would use a simple incremental system and give each new customer a numerical ID with a value one higher than the last. This, of course, is impossible on a distributed system such as ours, as each installation has no idea how many others there might be. In an ideal world, we would persuade all our customers to register with us, perhaps with the offer of a free sandwich, but short of bribing our customers, there is another, rather clever technique for generating genuinely unique identifiers on distributed systems.

The **Universally Unique Identifier** (UUID) is a method of creating unique values that is available as a `java.util`. There are several versions, some of which are based on namespaces, which are unique identifiers in themselves. The version we use here (version 4) uses a random number generator. It might be tempting to think this might produce duplicates, but the way the identifier is constructed means that there would have to be a download every second for twenty billion years before there was a serious risk of duplication, so for the purposes of our sandwich vendor, this system is probably adequate.

 There are many other features we could use here, such as adding a hit counter to preferences and using it to count how many times our app has been accessed over how many sandwiches we have sold or keeping a total of monies spent.

Welcoming a new user and adding an ID are things we only want to do the first time the application is run, so we will add both features at once. Here are the steps required to add welcome features and assign a unique user ID:

1. Add these two fields, setters, and getters to the `User` class:

```
private static boolean newUser;
private static String userId;


...

public boolean getNewUser() {
    return newUser;
}

public void setNewUser(boolean newUser) {
    User.newUser = newUser;
}

public String getUserId() {
    return userId;
}

public void setUserId(String userId) {
    User.userId = userId;
}
```

2. Add this code to the `loadPrefs()` method:

```
if (prefs.getBoolean("new_user", true)) {
    // Display welcome dialog
    // Add free credit for new users
    String uuid = UUID.randomUUID().toString();
    prefs.edit().putString("user_id", uuid);
    prefs.edit().putBoolean("new_user", false).apply();
}
```

Our app can now welcome and identify each and every user of our app. The beauty of using shared preferences to run code only the very first time an app is run, is that this method will ignore updates.

A somewhat simpler but less elegant solution to creating user IDs is to take the device's serial number, which can be achieved with something like this: `user.setId(`**Build.SERIAL**`.toString())`.

Summary

We have covered two quite separate topics in this chapter and covered both theoretical and practical subject matter. The composite pattern is incredibly useful and we saw how it could easily be used in the stead of other patterns, such as the builder.

Patterns are of no use if we do not have a handle on the more mechanical processes our software has to perform, such as file storage, and it should be clear that the list-like nature of data files, such as the shared preferences we worked with earlier, would be well-suited to builder patterns, and more complex data structures could be handled with composites.

In the next chapter, we will look at more non-immediate structures as we explore how to create services and post notifications to users when our application is not currently active. This will introduce observer patterns, which you will no doubt have encountered, in the form of listener methods.

9
Observing Patterns

In the last chapter, we looked at how to simplify interaction by allowing a user to store frequently used data, such as location and dietary preferences. This is only one way to make using an app as enjoyable as possible. Another valuable method is to provide the user with timely notifications.

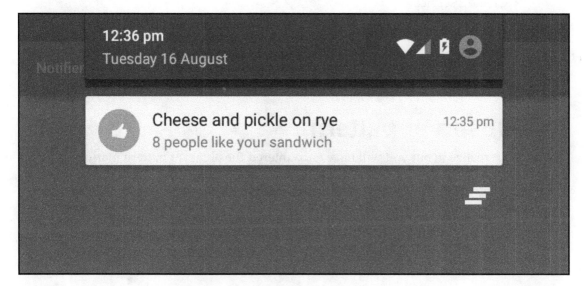

All mobile devices have provision for receiving notifications; usually these are delivered using the narrow status bar at the top of the screen, and Android is no exception. What makes this process interesting to us as developers, is that these notifications need to be delivered when our app may well not be in use. There is clearly no callback method for handling such an event in an activity, so we will have to look at background components such as **services** to trigger such events.

In terms of design patterns, there is one pattern that is almost purpose built for managing one to many relationships, the **observer pattern**. Although perfect for the delivery and reception of notifications, observer patterns occur everywhere in software design and you will no doubt have encountered the **Observer** and **Observed** Java utilities.

We will start the chapter by taking a close look at the observer pattern itself and then at how Android notifications are designed, built, and customized.

In this chapter, you will learn how to:

- Create an observer pattern
- Issue a notification
- Use Java observer utilities
- Apply a pending intent
- Configure privacy and priority settings
- Customize notifications
- Create a service

This chapter focuses largely on the observer pattern, and how it can be applied to managing notifications. The best place to begin is to take a look at the pattern itself, its purpose and its structure.

The Observer pattern

You may not realize it, but you will have encountered the observer pattern many times, as every click listener (and any other listener) is in fact an observer. The same applies to the icons and features of any desktop or GUI and these type of listener interfaces demonstrate very nicely the purpose of the observer pattern.

- The observer acts like a sentry, keeping watch for a particular event or state change in its subject, or subjects, and then reporting this information to interested parties.

As already mentioned, Java has its own observer utilities, and although these can be useful in some cases, the way Java handles inheritance and the simplistic nature of the pattern makes writing our own preferable. We will see how to use these built-in classes but in most of the examples here, we will build our own. This will also provide a deeper understanding of the pattern's workings.

Notifications must be used with caution, as few things can annoy a user more than unwanted messages. However, if used sparingly, notifications can provide an invaluable promotional tool. The secret lies in allowing users to opt in and out of various message streams, so that they only receive notifications that interest them.

Creating the pattern

Considering our sandwich maker app, there appears to be very little opportunity to issue notifications. One such use might be if we were to provide the option for customers to collect their sandwiches as well as having them delivered, then users may appreciate being notified when their sandwich is ready.

To effectively communicate between devices, we would need a central server with an associated application. We will not be able to cover that here but that will not stop us seeing how the pattern works and how notifications are posted.

We will begin by building a simple observer pattern, along with a basic notification manager to track and report the progress of an order.

To see how this is done, follow these steps:

1. At the heart of an observer pattern lies an interface for the subject and one for the observer.
2. The subject interface looks like this:

```
public interface Subject {

    void register(Observer o);
    void unregister(Observer o);
    boolean getReady();
    void setReady(boolean b);
}
```

3. This is the observer interface:

```
public interface Observer {

    String update();
}
```

4. Next, implement the subject as the sandwich being ordered, like so:

```java
public class Sandwich implements Subject {
    public boolean ready;

    // Maintain a list of observers
    private ArrayList<Observer> orders = new ArrayList<Observer>();

    @Override
    // Add a new observer
    public void register(Observer o) {
        orders.add(o);
    }

    @Override
    // Remove observer when order complete
    public void unregister(Observer o) {
        orders.remove(o);
    }

    @Override
    // Update all observers
    public void notifyObserver() {
        for (Observer order : orders) {
            order.update();
        }
    }

    @Override
    public boolean getReady() {
        return ready;
    }

    public void setReady(boolean ready) {
        this.ready = ready;
    }
}
```

5. Next, implement the observer interface like this:

```java
public class Order implements Observer {
    private Subject subject = null;

    public Order(Subject subject) {
        this.subject = subject;
    }

    @Override
```

```
public String update() {

    if (subject.getReady()) {

        // Stop receiving notifications
        subject.unregister(this);

        return "Your order is ready to collect";

    } else {
        return "Your sandwich will be ready very soon";
    }
}
}
```

This completes the pattern itself; its structure is quite simple as can be seen here:

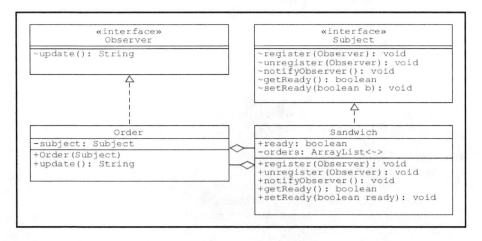

It is the subject that does all the work here. It keeps a list of all its observers and provides mechanisms for observers to subscribe and unsubscribe to its updates. In the previous example, we call `unregister()` during `update()` from the observer once the order is complete, as our listener will no longer be interested in this subject.

The `Observer` interface may seem too simple to be necessary, but it allows for loose coupling between the `Sandwich` and its observers, and this means we can modify either of them independently of the other.

Although we have only included one observer, it should be clear how the methods implemented in our subject allow for any number of individual orders and respond accordingly.

Adding a notification

The `order.update()` method provides the appropriate text for us to issue as a notification. To test the pattern and deliver notifications to the status bar, follow the steps here:

1. Begin by creating an XML layout with the following nested layout:

```xml
<LinearLayout
    ...
    android:layout_alignParentBottom="true"
    android:layout_centerHorizontal="true"
    android:gravity="end"
    android:orientation="horizontal">

    <Button
        android:id="@+id/action_save"
        style="?attr/borderlessButtonStyle"
        android:layout_width="wrap_content"
        android:layout_height="wrap_content"
        android:minWidth="64dp"
        android:onClick="onOrderClicked"
        android:padding="@dimen/action_padding"
        android:text="ORDER"
        android:textColor="@color/colorAccent"
        android:textSize="@dimen/action_textSize" />

    <Button
        android:id="@+id/action_update"
        ...
        android:onClick="onUpdateClicked"
        android:padding="@dimen/action_padding"
        android:text="UPDATE"
        ...
        />

</LinearLayout>
```

2. Open your Java activity and add these fields:

```java
Sandwich sandwich = new Sandwich();
Observer order = new Order(sandwich);

int notificationId = 1;
```

3. Add the method that listens for the order button to be clicked:

```
public void onOrderClicked(View view) {

    // Subscribe to notifications
    sandwich.register(order);
    sendNotification(order.update());
}
```

4. And one for the update button:

```
public void onUpdateClicked(View view) {

    // Mimic message from server
    sandwich.setReady(true);
    sendNotification(order.update());
}
```

5. Finally, add the `sendNotification()` method:

```
private void sendNotification(String message) {

    NotificationCompat.Builder builder =
            (NotificationCompat.Builder)
            new NotificationCompat.Builder(this)
                    .setSmallIcon(R.drawable.ic_stat_bun)
                    .setContentTitle("Sandwich Factory")
                    .setContentText(message);

    NotificationManager manager = (NotificationManager)
            getSystemService(NOTIFICATION_SERVICE);
    manager.notify(notificationId, builder.build());

    // Update notifications if needed
    notificationId += 1;
}
```

We can now run the code on a device or emulator:

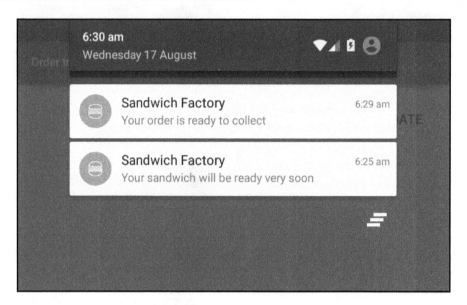

The code above, responsible for issuing the notifications, demonstrates the simplest possible way to post a notification, and the icon and two text fields are the minimum required for this.

 As this is only a demonstration and the observer pattern really does nothing more than simulate the server, it is important not to confuse this with the native notification API calls.

The use of a notification ID is worth noting. This is primarily used for updating notifications. Sending a notification with the same ID will update the previous message, and in the preceding situation, this is actually what we should have done, the incrementing of the ID here was done only to demonstrate how it can be used. To correct this, comment out the line and rerun the project so that only one message stream is generated.

There is a lot more we can and should do to make the most of this invaluable tool, such as have it perform actions and be delivered when our app is not active, and we will return to these issues in later sections, but for now it is worth taking a look at how Java provides its own utilities for implementing observer patterns.

Utility observers and observables

As mentioned earlier, Java provides its own observer utilities, the `java.util.observer` interface and the `java.util.observable` abstract class. These come equipped with methods for registering, un-registering, and notifying observers. The example from the previous section can be easily implemented using them, as can be seen by following these steps:

1. In this case, the subject is realized by extending the observable class, as can be seen here:

```
import java.util.Observable;

public class Sandwich extends Observable {
    private boolean ready;

    public Sandwich(boolean ready) {
        this.ready = ready;
    }

    public boolean getReady() {
        return ready;
    }

    public void setReady(boolean ready) {
        this.ready = ready;
        setChanged();
        notifyObservers();
    }
}
```

2. The `Order` class is an observer and therefore implements that interface, like this:

```
import java.util.Observable;
import java.util.Observer;

public class Order implements Observer {
    private String update;

    public String getUpdate() {
        return update;
    }

    @Override
    public void update(Observable observable, Object o) {
        Sandwich subject = (Sandwich) observable;
```

```
            if (subject.getReady()) {
                subject.deleteObserver(this);
                update = "Your order is ready to collect";

            } else {
                update = "Your sandwich will be ready very soon";
            }
        }
    }
```

3. The XML layout and `sendNotification()` method are exactly as before, and the only changes to the activity source code are as outlined here:

```
public class MainActivity extends AppCompatActivity {
    Sandwich sandwich = new Sandwich(false);
    Order order = new Order();
    private int id;

    @Override
    protected void onCreate(Bundle savedInstanceState)
        { ... }

    public void onOrderClicked(View view) {
        sandwich.addObserver(order);
        sandwich.setReady(true);
        sendNotification(order.getUpdate());
    }

    public void onUpdateClicked(View view) {
        sandwich.setReady(true);
        sendNotification(order.getUpdate());
    }

    private void sendNotification(String message)
        { ... }
}
```

As you can see, this code performs the same task as our previous example, and it is worth comparing the two listings. The Observer's `setChanged()` and `notifyObservers()` methods replaced the method we implemented in our custom version.

Which of these approaches you adopt for future observer patterns depends mostly on the particular circumstances. Generally, the use of the Java observable utils suits simple situations, and if you are unsure it is a good idea to start with this method, as you will soon see if a more flexible approach is required.

The examples covered above only introduce the observer pattern and notifications. The pattern demonstrated a very simple situation and to appreciate its full potential, we will need to apply it to a more complex situation. First though, we will take a look at how much more we can do with the notification system.

Notifications

Sending a simple string message to a user is the primary purpose of the notification system, but there is much more that can be done with it. First and foremost, a notification can be made to perform one or more actions; usually one of these will be to open the relevant application. It is also possible to create expanded notifications that can contain various media and are very useful for situations where there is too much information for a single line message, but we want to save the user the bother of having to open an app.

Since API 21, it has been possible to send heads-up notifications and notifications to the user's lock screen. This function is something that was taken from other mobile platforms and despite its apparent usefulness, it should be used with great caution. It barely requires pointing out that notifications should only contain pertinent and tine-related information. The rule of thumb is only issue a notification if the information cannot wait until the next time the user logs in. A good example of a valid notification might be *your sandwich has been delayed*, not *new range of cheeses coming soon*.

Along with the risk of pestering the user, the lock screen notification contains another danger. Messages displayed on a locked device are to all intents and purposes public. Anyone passing a phone left on a desk can see the content. Now although most people would not mind their boss seeing what type of sandwich they like, there will no doubt be apps you will write that will contain more sensitive material and fortunately the API provides programmable privacy settings.

Regardless of the caution that needs applying, the full range of notification functionality is still well worth becoming acquainted with, starting with having a notification actually do something.

Setting an intent

As with the starting of an activity or any other top level app component, the intent provides our route from notification to action. In most cases, we want to use the notification to start an activity, and this is what we will do here.

Users of mobile devices want to be able to move between activities and applications easily and swiftly. As a user navigates between apps, the system keeps track of the order which it stores in a back-stack. This is usually sufficient, but when a user is drawn away from an app by a notification, then pressing the back button will not return them to the app they had previously been engaged in. This is likely to annoy users but is fortunately easily avoided by creating an artificial back-stack.

Creating our own back-stack is nowhere as difficult as it might sound as the following example demonstrates. It is in fact so simple that this example also details how to include a few other notification features, such as a more elaborate icon and ticker text that will scroll along the status bar when the notification is first delivered.

Follow these steps to see how this is achieved:

1. Open the project we worked on earlier and create a new activity class, like so:

```
public class UserProfile extends AppCompatActivity {

    @Override
    protected void onCreate(Bundle savedInstanceState) {
        super.onCreate(savedInstanceState);
        setContentView(R.layout.activity_profile);
    }
}
```

2. Next we will need a layout file to match the content view set in the `onCreate()` method previously. This can be left empty, save for a root layout.
3. Now add the following lines to the top of the `sendNotification()` method in your main activity:

```
Intent profileIntent = new Intent(this, UserProfile.class);

TaskStackBuilder stackBuilder = TaskStackBuilder.create(this);
stackBuilder.addParentStack(UserProfile.class);
stackBuilder.addNextIntent(profileIntent);

PendingIntent pendingIntent = stackBuilder.getPendingIntent(0,
        PendingIntent.FLAG_UPDATE_CURRENT);
```

4. Append the notification builder with these settings:

```
.setAutoCancel(true)
.setTicker("the best sandwiches in town")
.setLargeIcon(BitmapFactory.decodeResource(getResources(),
        R.drawable.ic_sandwich))
.setContentIntent(pendingIntent);
```

5. Finally, include the new activity in the manifest file:

```
<activity android:name="com.example.kyle.ordertracker.UserProfile">

    <intent-filter>
        <action android:name="android.intent.action.DEFAULT" />
    </intent-filter>

</activity>
```

The effects of these changes are obvious:

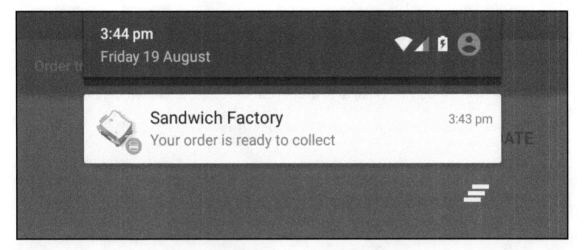

Comment out the lines generating the back-stack and open the notification when using another app to see how it maintains intuitive navigation. The `setAutoCancel()` call means that the status bar icon is dismissed when the notification is followed.

Generally, we want a user to open our app from a notification, but from a user's perspective it is important to get things done with the least possible effort and if they can garner the same information without having to open another app, then that is a good thing. This is where the expanded notification comes in.

Customizing and configuring notifications

The expanded notification was introduced with API 16. The providing of a larger, more flexible content area brought it in line with other mobile platforms. There are three styles of expanded notification: text, image, and list. The following steps demonstrate how to implement each of them:

1. The following project can either be modified from the one we used earlier, or begun from scratch.
2. Edit the main layout file so that it contains three buttons with the following observer methods:

```
android:onClick="onTextClicked"
android:onClick="onPictureClicked"
android:onClick="onInboxClicked"
```

3. Make the following changes to the `sendNotification()` method:

```
private void sendNotification(NotificationCompat.Style style) {

    . . .

    NotificationCompat.Builder builder = (NotificationCompat.Builder) new
NotificationCompat.Builder(this)

        .setStyle(style)

        . . .

    manager.notify(id, builder.build());
}
```

4. Now create the three style methods. First the big text style:

```
public void onTextClicked(View view) {
    NotificationCompat.BigTextStyle bigTextStyle = new
NotificationCompat.BigTextStyle();

    bigTextStyle.setBigContentTitle("Congratulations!");
    bigTextStyle.setSummaryText("Your tenth sandwich is on us");
    bigTextStyle.bigText(getString(R.string.long_text));

    id = 1;
    sendNotification(bigTextStyle);
}
```

5. The big picture style requires these settings:

```
public void onPictureClicked(View view) {
    NotificationCompat.BigPictureStyle bigPictureStyle = new
NotificationCompat.BigPictureStyle();

    bigPictureStyle.setBigContentTitle("Congratulations!");
    bigPictureStyle.setSummaryText("Your tenth sandwich is on us");
    bigPictureStyle.bigPicture(BitmapFactory.decodeResource(getResources(),
R.drawable.big_picture));

    id = 2;
    sendNotification(bigPictureStyle);
}
```

6. Finally add the list, or inbox, style, like so:

```
public void onInboxClicked(View view) {
    NotificationCompat.InboxStyle inboxStyle = new
NotificationCompat.InboxStyle();

    inboxStyle.setBigContentTitle("This weeks most popular sandwiches");
    inboxStyle.setSummaryText("As voted by you");

    String[] list = {
            "Cheese and pickle",
            ...
    };

    for (String l : list) {
        inboxStyle.addLine(l);
    }

    id = 3;
    sendNotification(inboxStyle);
}
```

These notifications can now be tested on a device or AVD:

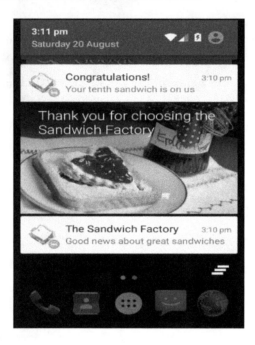

The most recent notification will always be expanded and others can be expanded by swiping down on them. As with most material lists, notifications can be dismissed by swiping them horizontally.

These features provide us with a lot of flexibility when it comes to designing notifications and if we want to do more, we can even customize them. This can be done very simply by passing an XML layout to our builder. To do this, we need the RemoteViews class, which is a form of layout inflater. Create a layout and then include the following line in your code to instantiate it:

```
RemoteViews expandedView = new RemoteViews(this.getPackageName(),
R.layout.notification);
```

Then pass this to the builder with:

```
builder.setContent(expandedView);
```

In terms of implementing Android notifications, all we need to cover is how to issue heads-up notifications and lock screen notifications. This is more a matter of setting priorities and user permissions and settings than coding.

Visibility and priority

Where and how some notifications appear is often down to two related properties, privacy and importance. These are applied using metadata constants and can also include categories such as *alarm* and *promo*, which the system can use to sort and filter multiple notifications.

When it comes to delivering a notification to a user's lock screen, it is not only how we set the metadata, it also depends on the user's security setup. To view these notifications, the user will have had to select a secure lock like a PIN or gesture and then choose one of the following options from the **Security | Notifications** settings:

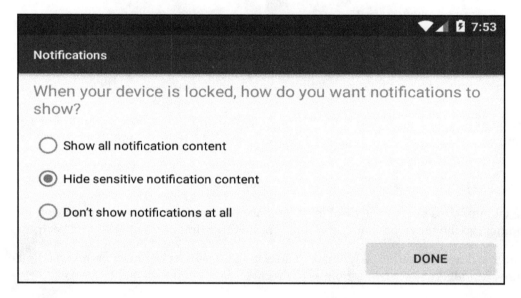

Providing the user has these settings, our notifications will be delivered to a user's lock screen. To protect the user's privacy, we can set notification visibility with our builder. There are three values for this:

- `VISIBILITY_PUBLIC` – The entire notification is displayed
- `VISIBILITY_PRIVATE` – Title and icon are displayed but content is hidden
- `VISIBILTY_SECRET` – Nothing is displayed at all

To implement any of these settings use a line like this:

```
builder.setVisibility(NotificationCompat.VISIBILITY_PUBLIC)
```

Heads up displays alert the user to their importance by appearing as a basic (collapsed) notification at the top of the screen for five seconds, before reverting to a status bar icon. They should only be used for information that requires the user's immediate attention. This is controlled using priority metadata.

By default, the priority of each notification is PRIORITY_DEFAULT. The five possible values are:

- `PRIORITY_MIN` = -2
- `PRIORITY_LOW` = -1
- `PRIORITY_DEFAULT` = 0
- `PRIORITY_HIGH` = 1
- `PRIORITY_MAX` = 2

These too can be set by the builder, for example:

```
builder.setPriority(NotificationCompat.PRIORITY_MAX)
```

Any value greater than DEFAULT will trigger a heads-up notification, providing either sound or vibration are also triggered. This too can be added by our builder and would take the following form:

```
builder.setVibrate(new long[]{500, 500, 500})
```

The vibrator class takes an array of longs and applies these as millisecond bursts of vibration, so the previous example would vibrate three times for half a second each.

Including device vibrations anywhere in an app requires user permissions on installment. These are added to the manifest file as a direct child of the root element, like so:

```
<manifest xmlns:android="http://schemas.android.com/apk/res/android"
    package="com.example.yourapp">

    <uses-permission
        android:name="android.permission.VIBRATE" />

    <application

        . . .

    </application>

</manifest>
```

There is little more we need to know about displaying and configuring notifications. However, so far we have been issuing notifications from within the app itself rather than remotely as they would in the wild.

Services

Services are top-level application components such as activities. Their purpose is to manage long-running background tasks such as playing audio or triggering reminders or other scheduled events. Services do not require a UI, but in other respects they are similar to activities and have a similar life cycle with associated callback methods that we can use to intercept key events.

Although all services start out the same, they basically fall into two categories, bound and unbound. A service that is bound to an activity will continue to run until instructed otherwise or the binding activity is stopped. An unbound service, on the other hand, will continue to run regardless of whether the calling activity is active or not. In both these cases, the service will often be responsible for switching itself off once it has completed its allotted task.

The following example demonstrates how to create a service that will set a reminder. The service will then either post a notification after a set delay or be canceled by user action. To see how this is done, follow these steps:

1. Start by creating a layout. This will need two buttons:

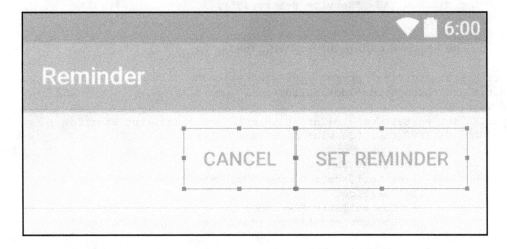

2. Include onClick attributes in both buttons:

```
android:onClick="onReminderClicked"
android:onClick="onCancelClicked"
```

3. Create a new class to extend Service:

```
public class Reminder extends Service
```

4. The `onBind()` method will be insisted upon, but we will not need it, so it can be left like this:

```
@Override
public IBinder onBind(Intent intent) {
    return null;
}
```

5. We will not be using the `onCreate()` or `onDestroy()` methods either but it is always useful to see how background activities are behaving, so complete the methods like so:

```
@Override
public void onCreate() {
    Log.d(DEBUG_TAG, "Service created");
}

@Override
public void onDestroy() {
    Log.d(DEBUG_TAG, "Service destroyed");
}
```

6. The class will require the following fields:

```
private static final String DEBUG_TAG = "tag";
NotificationCompat.Builder builder;
@Override
public int onStartCommand(Intent intent, int flags, int startId) {
    Log.d(DEBUG_TAG, "Service StartCommand");

    // Build notification
    builder = new NotificationCompat.Builder(this)
            .setSmallIcon(R.drawable.ic_bun)
            .setContentTitle("Reminder")
            .setContentText("Your sandwich is ready to collect");

    // Issue timed notification in separate thread
    new Thread(new Runnable() {
```

```
@Override
public void run() {
    Timer timer = new Timer();
    timer.schedule(new TimerTask() {

        @Override
        public void run() {
            NotificationManager manager = (NotificationManager)
                    getSystemService(NOTIFICATION_SERVICE);
            manager.notify(0, builder.build());
            cancel();
        }

    // Set ten minute delay
    }, 1000 * 60 * 10);
    // Destroy service after first use
    stopSelf();
}

}).start();

return Service.START_STICKY;
}
```

7. Add the service to the manifest file alongside your activities, like this:

```
<service
    android:name=".Reminder" />
```

8. Finally, open your main Java activity and complete these two button listeners:

```
public void onReminderClicked(View view) {
    Intent intent = new Intent(MainActivity.this, Reminder.class);
    startService(intent);
}

public void onCancelClicked(View view) {
    Intent intent = new Intent(MainActivity.this, Reminder.class);
    stopService(intent);
}
```

The previous code demonstrates how to run code in the background using a service. In many apps, this is an essential feature in many apps. Our only real consideration is to ensure that all our services are correctly disposed of when they are no longer needed, as services are particularly susceptible to memory leakage.

Summary

In this chapter, we have seen how the observer pattern can be used as a tool to manage the delivery of user notifications, as well as keeping track of many other events and responding accordingly. We began by looking at the pattern itself and then at the Android notification APIs, which despite using the system-controlled status bar and notification drawer, allow us a great deal of freedom when it comes to designing notifications that suit the purpose and look of our apps.

In the next chapter, we will take this and other patterns and see how we can extend extant Android components and have them apply our design patterns directly. We will then see how this can help us when it comes to developing for form factors other than phones and tablets.

10
Behavioral Patterns

So far in this book, we have looked closely at many of the most significant creational and structural design patterns. This has given us the power to construct all manner of architectures, but to perform the tasks we require, these structures need to be able to communicate, between their own elements and with other structures.

Behavioral patterns were designed for modeling many of the general development problems we encounter on a regular basis, such as responding to a change in state of a particular object or adapting behavior to accord with a hardware change. We have already encountered one, in the last chapter with the observer, and here we will look further into some of the most useful behavioral patterns.

Behavioral patterns are far more adaptable in terms of the variety of tasks they can perform than creational and behavioral patterns, and although this flexibility is great, it can also complicate matters when it comes to selecting the best possible pattern, as there will often be two or three possible candidates for a given task. It is a good idea to take a look at several of these patterns together and see how understanding these sometimes subtle differences can help us apply behavioral patterns efficiently.

In this chapter, you will learn how to:

- Create a template pattern
- Add specialization layers to the pattern
- Apply a strategy pattern
- Build and use a visitor pattern
- Create a state machine

The generalized nature of these patterns means they can be applied in a huge number of various situations, but a good example of the type of task they can perform would be a click or touch listener and of course, the observer patterns of the previous chapter. Another common feature seen in many behavioral patterns is the use of abstract classes to create generalized algorithms, as we will see later in the chapter with the **visitor** and **strategy patterns** and in particular the **template pattern**, which we shall explore now.

The template pattern

Even if you are entirely new to design patterns, you will be familiar with how the template pattern works, as it employs abstract classes and methods to form a generalized (template) solution that can be used to create specialized subclasses in precisely the way that abstraction is intended to be used in OOP.

At its very simplest, the template pattern is nothing more than a generalization in the form of an abstract class with at least one concrete realization. For example, a template might define an empty layout and its realizations then control the content. One big advantage of this approach is that common elements and shared logic need only be defined in the base class, meaning that we only need to write code where our realizations differ from each other.

Template patterns can become even more powerful and flexible if another layer of abstraction is added in the form of specializations of the base class. These can then be used as sub-categories of their parent classes and treated similarly. Before exploring these multi-layered patterns, we will take a look at the simplest example of a base template that provides the properties and logic to produce different outputs in accordance with their concrete realizations.

Generally speaking, template patterns work on an algorithm or any set of procedures that can be broken down into steps. This template method is defined in the base class and made specific with realizations.

The best way to see this is by way of example. Here we will imagine a simple news feed app with a generalized *story* template with *news* and *sport* realizations. Follow these steps to create this pattern:

1. Start a new project and create a main layout based on the following component tree:

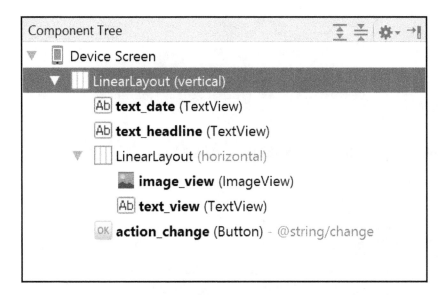

2. Create a new abstract class called Story, as our generalization, like so:

```
abstract class Story {
    public String source;

    // Template skeleton algorithm
    public void publish(Context context) {
        init(context);
        setDate(context);
        setTitle(context);
        setImage(context);
        setText(context);
    }

    // Placeholder methods
    protected abstract void init(Context context);

    protected abstract void setTitle(Context context);

    protected abstract void setImage(Context context);
```

```
            protected abstract void setText(Context context);

            // Calculate date as a common property
            protected void setDate(Context context) {
                Calendar calendar = new GregorianCalendar();
                SimpleDateFormat format =
                    new SimpleDateFormat("MMMM d");

                format.setTimeZone(calendar.getTimeZone());

                TextView textDate = (TextView)
                    ((Activity) context)
                    .findViewById(R.id.text_date);
                textDate.setText(format.format(calendar.getTime()));
            }
        }
```

3. Now, extend this to create the News class, like so:

```
        public class News extends Story {
            TextView textHeadline;
            TextView textView;
            ImageView imageView;

            @Override
            protected void init(Context context) {
                source = "NEWS";
                textHeadline = (TextView) ((Activity)
context).findViewById(R.id.text_headline);
                textView = (TextView) ((Activity)
context).findViewById(R.id.text_view);
                imageView = (ImageView) ((Activity)
context).findViewById(R.id.image_view);
            }

            @Override
            protected void setTitle(Context context) {
                ((Activity)
context).setTitle(context.getString(R.string.news_title));
            }

            @Override
            protected void setImage(Context context) {
                imageView.setImageResource(R.drawable.news);
            }

            @Override
            protected void setText(Context context) {
```

```
                textHeadline.setText(R.string.news_headline);
                textView.setText(R.string.news_content);
            }
        }
```

4. The `Sport` realization is the same, but with the following exceptions:

```
        public class Sport extends Story {
            ...

            @Override
            protected void init(Context context) {
                source = "NEWS";
                ...
            }

            @Override
                protected void setTitle(Context context) {
                ((Activity)
context).setTitle(context.getString(R.string.sport_title));
            }

            @Override
            protected void setImage(Context context) {
                imageView.setImageResource(R.drawable.sport);
            }

            @Override
            protected void setText(Context context) {
                textHeadline.setText(R.string.sport_headline);
                textView.setText(R.string.sport_content);
            }
        }
```

5. Finally, add these lines to the main activity:

```
public class MainActivity
    extends AppCompatActivity
    implements View.OnClickListener {

    String source = "NEWS";
    Story story = new News();

    @Override
    protected void onCreate(Bundle savedInstanceState) {
        ...

    Button button = (Button)
```

```
            findViewById(R.id.action_change);
        button.setOnClickListener(this);

        story.publish(this);
    }

    @Override
    public void onClick(View view) {

        if (story.source == "NEWS") {
            story = new Sport();

        } else {
            story = new News();
        }

        story.publish(this);
    }
}
```

Running this code on a real or virtual device allows us to toggle between the two realizations of our `Story` template:

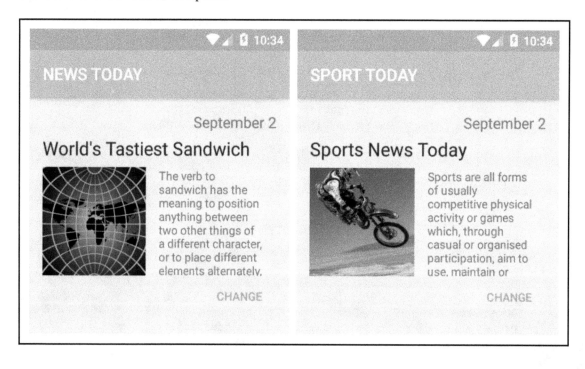

This template example is nice and simple and familiar but nevertheless the template can be applied in numerous situations and provides a very convenient method to organize code, in particular when there are many derived classes to be defined. The class diagram is as straightforward as the code:

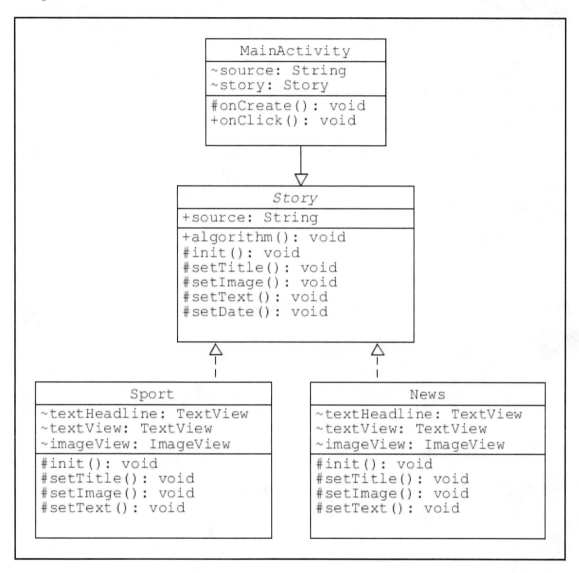

Extending templates

The previous pattern is very useful when the individual realizations are very similar to each other, but there are often cases when we want to model objects that are similar enough to each other to warrant shared code but nevertheless possess different types or number of properties. A good example might be a database for a reading library. We could create a base class called *ReadingMaterial* and with the right properties, this could be used to cover almost any book, regardless of genre or content or age. If, however, we wanted to include magazines and periodicals, we might find our model failing to represent the multiple nature of such periodicals. In this situation, we could create an entirely new base class or alternatively create new and specialized abstract classes that extend the base class but are also extendable themselves.

We will use this above example to demonstrate this, more functional, template pattern. This model now has three layers, generalizations, specializations, and realizations. As it is the structure of the pattern that matters here, we will save time and use the debugger to output our realized objects. To see how this would be put in practice, follow these steps:

1. Begin by creating an abstract, base class, like so:

```java
abstract class ReadingMaterial {

    // Generalization
    private static final String DEBUG_TAG = "tag";
    Document doc;

    // Standardized skeleton algorithm
    public void fetchDocument() {
        init();
        title();
        genre();
        id();
        date();
        edition();
    }

    // placeholder functions
    protected abstract void id();

    protected abstract void date();

    // Common functions
    private void init() {
        doc = new Document();
    }
```

```
    private void title() {
        Log.d(DEBUG_TAG,"Title : "+doc.title);
    }

    private void genre() {
        Log.d(DEBUG_TAG, doc.genre);
    }

    protected void edition() {
        Log.d(DEBUG_TAG, doc.edition);
    }
}
```

2. Next, another abstract class for the book category:

```
abstract class Book extends ReadingMaterial {

    // Specialization
    private static final String DEBUG_TAG = "tag";

    // Override implemented base method
    @Override
    public void fetchDocument() {
        super.fetchDocument();
        author();
        rating();
    }

    // Implement placeholder methods
    @Override
    protected void id() {
        Log.d(DEBUG_TAG, "ISBN : " + doc.id);
    }

    @Override
    protected void date() {
        Log.d(DEBUG_TAG, doc.date);
    }

    private void author() {
        Log.d(DEBUG_TAG, doc.author);
    }

    // Include specialization placeholder methods
    protected abstract void rating();
}
```

3. The `Magazine` class should look like this:

```
abstract class Magazine extends ReadingMaterial {

    //Specialization
    private static final String DEBUG_TAG = "tag";

    // Implement placeholder methods
    @Override
    protected void id() {
        Log.d(DEBUG_TAG, "ISSN : " + doc.id);
    }

    @Override
    protected void edition() {
        Log.d(DEBUG_TAG, doc.period);
    }

    // Pass placeholder on to realization
    protected abstract void date();
}
```

4. Now we can create the concrete realization classes. First the book class:

```
public class SelectedBook extends Book {
    // Realization
    private static final String DEBUG_TAG = "tag";

    // Implement specialization placeholders
    @Override
    protected void rating() {
        Log.d(DEBUG_TAG, "4 stars");
    }
}
```

5. Followed by the magazine class:

```
public class SelectedMagazine extends Magazine {
    // Realization
    private static final String DEBUG_TAG = "tag";

    // Implement placeholder method only once instance created
    @Override
    protected void date() {
        Calendar calendar = new GregorianCalendar();
        SimpleDateFormat format = new SimpleDateFormat("MM-d-yyyy");
        format.setTimeZone(calendar.getTimeZone());
        Log.d(DEBUG_TAG, format.format(calendar.getTime()));
```

```
        }
}
```

6. Create a POJO to use as dummy data, like so:

```
public class Document {
    String title;
    String genre;
    String id;
    String date;
    String author;
    String edition;
    String period;

    public Document() {
        this.title = "The Art of Sandwiches";
        this.genre = "Non fiction";
        this.id = "1-23456-789-0";
        this.date = "06-19-1993";
        this.author = "J Bloggs";
        this.edition = "2nd edition";
        this.period = "Weekly";
    }
}
```

7. This pattern can now be tested with code like the following in the main activity:

```
// Print book
ReadingMaterial document = new SelectedBook();
document.fetchDocument();
// Print magazine
ReadingMaterial document = new SelectedMagazine();
document.fetchDocument();
```

By changing the dummy document code, any realization can be tested, and will produce output along these lines:

```
D/tag: The Art of Sandwiches
D/tag: Non fiction
D/tag: ISBN : 1-23456-789-0
D/tag: 06-19-1963
D/tag: 2nd edition
D/tag: J Bloggs
D/tag: 4 stars
D/tag: Sandwich Weekly
D/tag: Healthy Living
D/tag: ISSN : 1-23456-789-0
D/tag: 09-3-2016
D/tag: Weekly
```

The previous example is short and simple, but it demonstrates each of the features that make the pattern so useful and versatile, as detailed in this list:

- Base classes provide standardized skeleton definitions and code, as demonstrated by the `fetchDocument()` method
- Code that realizations have in common is defined in base classes, for example `title()` and `genre()`
- Placeholders are defined in base classes for specialized implementations, as seen by the way `date()` is managed
- Derived classes can override placeholder methods and implemented methods; see `rating()`
- Derived classes can call back to base classes with `super`, as seen with the `fetchDocument()` method in the `Book` class

Although the template pattern may seem complex at first, the fact that so many elements are shared, means that well thought out generalizations and specializations can lead to very simple and clear code in the concrete classes themselves, something we will be grateful for when dealing with more than just one or two template realizations. This concentration of code defined in the abstract classes can be seen very clearly in the pattern's class diagram, where the derived classes contain only the code that pertains to it alone:

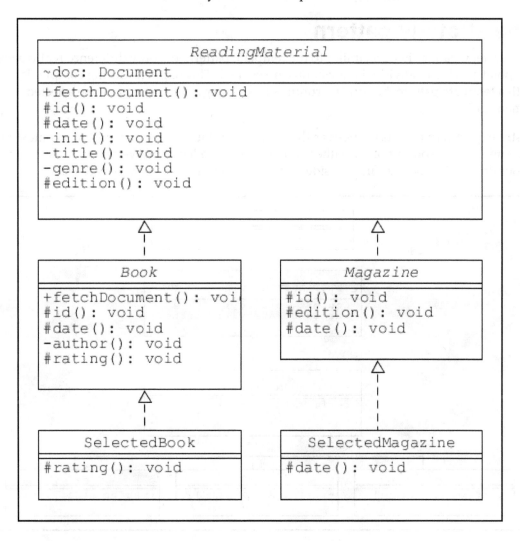

As mentioned at the top of the chapter, there are often more than one behavioral pattern that can be used in a given situation and the template pattern we discussed previously, along with the strategy, visitor, and state patterns, all fit into this category, as all of them derive specialized cases from generalized outlines. Each of these patterns deserves exploring in a little detail.

The strategy pattern

The strategy pattern is very similar to the template pattern, the only difference really being the point at which individual realizations are created. This happens during compilation with a template pattern, but during runtime in a strategy pattern, and can be selected dynamically.

A strategy pattern reflects changes as they happen and its output depends on a context in the same way the output of a weather app depends on a location. We can use this scenario in our demonstration, but first consider this class diagram of a strategy pattern:

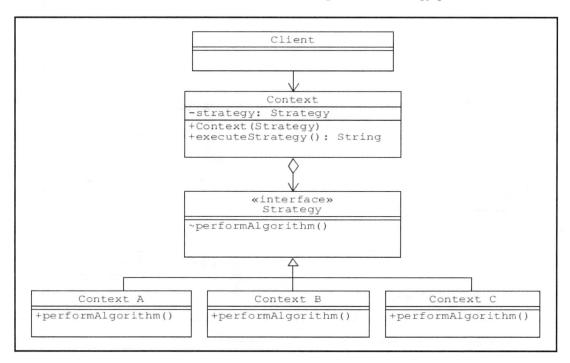

This can easily be realized using the weather example. Open a new project and follow these steps to see how:

1. Start with strategy interface; it looks like this:

```
public interface Strategy {

    String reportWeather();
}
```

2. Create several concrete implementations along the lines of the class here:

```
public class London implements Strategy {

    @Override
    public String reportWeather() {
        return "Constant drizzle";
    }
}
```

3. Next, create the context class, which here is the location:

```
public class Location {
    private Strategy strategy;

    public Location(Strategy strategy) {
        this.strategy = strategy;
    }

    public void executeStrategy(Context context) {
        TextView textView=(TextView)
                ((Activity)context)
                .findViewById(R.id.text_view);
        textView.setText(strategy.reportWeather());
    }
}
```

4. By mocking up the location with a string value, we can test the pattern with client code like this:

```
Location context;
String location = "London";

switch (location) {
    case "London":
        context = new Location(new London());
        break;
```

```
        case "Glasgow":
            context = new Location(new Glasgow());
            break;
        default:
            context = new Location(new Paris());
            break;
    }

    context.executeStrategy(this);
```

As this example demonstrates, strategy patterns, although similar to templates, are used for different tasks as they are applied at different occasions, runtime and compile time respectively.

As well as applying our own templates and strategies, most platforms apply their own as part of the system. A good example of a strategy pattern at work in the Android framework can be seen every time a device is rotated and templates are applied to install layouts for different devices. We will take a closer look at this shortly, but first there are two other patterns we need to examine.

The visitor pattern

Just like the template and strategy patterns, the visitor pattern is flexible enough to perform any of the tasks we have so far considered, and as with other behavioral patterns, the trick lies in applying the right pattern to the right problem. The term *visitor* is perhaps not as self-explanatory as *template* or *strategy*.

The visitor pattern is designed so that a client can apply a process to a collection of unrelated objects, without having to worry about their differences. A good real-world example would be a trip to a supermarket where we might buy tinned products that have a bar code that can be scanned as well as fresh items that need to be weighed. This difference need not concern us in a supermarket as the cashier will take care of all this for us. In this case, the cashier is acting as the visitor, making all the necessary decisions regarding how to process individual items, leaving us (the client) only having to consider the final bill.

This does not really accord with our intuitive understanding of the word *visitor*, but from a design pattern point of view this is what it means. Another real-world example would be if we wished to travel across town. In this example, we might choose between a taxi or a bus. In both cases, we only concern ourselves with the final destination (and perhaps the cost) leaving the driver/visitor to negotiate the details of the actual route.

Follow these steps to see how a visitor pattern can be implemented to model the supermarket scenario outlined previously:

1. Start a new Android project and add the following interface to define shopping items, like so:

```
public interface Item {

    int accept(Visitor visitor);
}
```

2. Next, create two item examples. First one for canned foods:

```
public class CannedFood implements Item {
    private int cost;
    private String name;

    public CannedFood(int cost, String name) {
        this.cost = cost;
        this.name = name;
    }

    public int getCost() {
        return cost;
    }

    public String getName() {
        return name;
    }

    @Override
    public int accept(Visitor visitor) {
        return visitor.visit(this);
    }
}
```

3. Next, add the fresh foods item class:

```
public class FreshFood implements Item {
    private int costPerKilo;
    private int weight;
    private String name;

    public FreshFood(int cost, int weight, String name) {
        this.costPerKilo = cost;
        this.weight = weight;
        this.name = name;
```

```
    }

    public int getCostPerKilo() {
        return costPerKilo;
    }

    public int getWeight() {
        return weight;
    }

    public String getName() {
        return name;
    }

    @Override
    public int accept(Visitor visitor) {
        return visitor.visit(this);
    }
}
```

4. Now we can add the visitor interface itself, like this:

```
public interface Visitor {

    int visit(FreshFood freshFood);
    int visit(CannedFood cannedFood);
}
```

5. This can then be implemented as the following `Checkout` class:

```
public class Checkout implements Visitor {
    private static final String DEBUG_TAG = "tag";

    @Override
    public int visit(CannedFood cannedFood) {
        int cost = cannedFood.getCost();
        String name = cannedFood.getName();
        Log.d(DEBUG_TAG, "Canned " + name + " : " + cost + "c");
        return cost;
    }

    @Override
    public int visit(FreshFood freshFood) {
        int cost = freshFood.getCostPerKilo() * freshFood.getWeight();
        String name = freshFood.getName();
        Log.d(DEBUG_TAG, "Fresh " + name + " : " + cost + "c");
        return cost;
    }
```

}

6. We can now see how the pattern allows us to write clean client code, like so:

```
public class MainActivity extends AppCompatActivity {
    private static final String DEBUG_TAG = "tag";

    private int totalCost(Item[] items) {
        Visitor visitor = new Checkout();
        int total = 0;
        for (Item item : items) {
            System.out.println();
            total += item.accept(visitor);
        }
        return total;
    }

    @Override
    protected void onCreate(Bundle savedInstanceState) {
        super.onCreate(savedInstanceState);
        setContentView(R.layout.activity_main);

        Item[] items = new Item[]{
                new CannedFood(65, "Tomato soup"),
                new FreshFood(60, 2, "Bananas"),
                new CannedFood(45, "Baked beans"),
                new FreshFood(45, 3, "Apples")};

        int total = totalCost(items);
        Log.d(DEBUG_TAG, "Total cost : " + total + "c");
    }
}
```

This should then produce an output like this:

```
D/tag: Canned Tomato soup : 65c
D/tag: Fresh Bananas : 120c
D/tag: Canned Baked beans : 45c
D/tag: Fresh Apples : 135c
D/tag: Total cost : 365
```

The visitor pattern has two particular strengths. The first is that it saves us having to use convoluted conditional nesting to differentiate between item types. The second, more significant, strength lies in how visitor and visited are kept separate, meaning that new item types can be added and altered without making any changes to the client at all. To see how this is done, just add the following code:

1. Open and edit the `Visitor` interface so that it has the extra line highlighted here:

```
public interface Visitor {

    int visit(FreshFood freshFood);
    int visit(CannedFood cannedFood);

    int visit(SpecialOffer specialOffer);
}
```

2. Create a `SpecialOffer` class, like so:

```
public class SpecialOffer implements Item {
    private int baseCost;
    private int quantity;
    private String name;

    public SpecialOffer(int cost,
                        int quantity,
                        String name) {
        this.baseCost = cost;
        this.quantity = quantity;
        this.name = name;
    }

    public int getBaseCost() {
        return baseCost;
    }

    public int getQuantity() {
        return quantity;
    }

    public String getName() {
        return name;
    }

    @Override
    public int accept(Visitor visitor) {
        return visitor.visit(this);
    }
```

```
}
```

   3. Overload the `visit()` method in the `Checkout` visitor class like so:

```
@Override
public int visit(SpecialOffer specialOffer) {

    String name = specialOffer.getName();
    int cost = specialOffer.getBaseCost();
    int number = specialOffer.getQuantity();
    cost *= number;

    if (number > 1) {
        cost = cost / 2;
    }

    Log.d(DEBUG_TAG, "Special offer" + name + " : " + cost + "c");
    return cost;
}
```

As this demonstrates, the visitor pattern can be extended to manage any number of items and any number of different solutions. Visitors can be used one at a time, or as part of a chain of processes, and are often used when importing files with different formats.

All the behavioral patterns we have looked at in this chapter have a very wide scope and can be used to solve an enormous variety of software design problems. There is one pattern, however, that has a wider scope even than these, the state design pattern or machine.

The state pattern

The state pattern is without doubt the most flexible of all the behavioral patterns. The pattern demonstrates how we can implement **finite state machines** in our code. State machines were an invention of the mathematician Alan Turing, who used them to realize all-purpose computers and prove that any mathematically computable process can be performed mechanically. In short, state machines can be used to perform any task we choose.

The mechanics of the state design pattern are simple and elegant. At any point in the life-cycle of a finite state machine, the pattern is aware of its own internal state and the current external state, or input. Based on these two properties, the machine will then produce an output (which can be none) and change its own internal state (which can be the same). Believe it or not, very sophisticated algorithms can be realized with properly configured finite state machines.

A traditional way of demonstrating the state pattern is with the example of a coin operated turnstile of the type that might be found at a sports stadium or funfair. This has two possible states, locked and unlocked, and takes two forms of input, a coin and a physical push.

To see how this can be modeled, follow these steps:

1. Start a new Android project and build a layout along the lines of the one here:

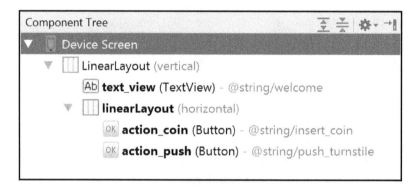

2. Add the following interface:

```
public interface State {

    void execute(Context context, String input);
}
```

3. Next comes the `Locked` state:

```
public class Locked implements State {

    @Override
    public void execute(Context context, String input) {

        if (Objects.equals(input, "coin")) {
            Output.setOutput("Please push");
            context.setState(new Unlocked());

        } else {
            Output.setOutput("Insert coin");
        }
    }
}
```

4. Followed by the `Unlocked` state:

```java
public class Unlocked implements State {

    @Override
    public void execute(Context context, String input) {

        if (Objects.equals(input, "coin")) {
            Output.setOutput("You have already paid");

        } else {
            Output.setOutput("Thank you");
            context.setState(new Locked());
        }
    }
}
```

5. Create the following singleton to hold the output string:

```java
public class Output {
    private static String output;

    public static String getOutput() {
        return output;
    }

    public static void setOutput(String o) {
        output = o;
    }
}
```

6. Next add the `Context` class, like this:

```java
public class Context {
    private State state;

    public Context() {
        setState(new Locked());
    }

    public void setState(State state) {
        this.state = state;
    }

    public void execute(String input) {
        state.execute(this, input);
    }
}
```

7. Finally edit the main activity to match this code:

```
public class MainActivity extends AppCompatActivity implements
View.OnClickListener {
    TextView textView;
    Button buttonCoin;
    Button buttonPush;

    Context context = new Context();

    @Override
    protected void onCreate(Bundle savedInstanceState) {
        super.onCreate(savedInstanceState);
        setContentView(R.layout.activity_main);

        textView = (TextView) findViewById(R.id.text_view);

        buttonCoin = (Button) findViewById(R.id.action_coin);
        buttonPush = (Button) findViewById(R.id.action_push);
        buttonCoin.setOnClickListener(this);
        buttonPush.setOnClickListener(this);
    }

    @Override
    public void onClick(View view) {

        switch (view.getId()) {

            case R.id.action_coin:
                context.execute("coin");
                break;

            case R.id.action_push:
                context.execute("push");
                break;
        }

        textView.setText(Output.getOutput());
    }
}
```

This example may be simple, but it demonstrates perfectly just how powerful this pattern is. It is easy to see how the same scheme could be expanded to model more complex locking systems, and finite state machines are often used to implement combination locks. As mentioned earlier, the state pattern can be used to model anything that can be mathematically modeled. The previous example is easily tested and easy to expand:

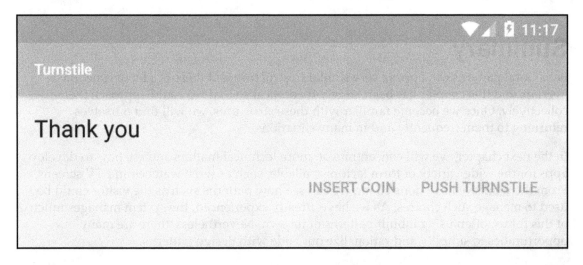

The true beauty of the state pattern lays in not just how incredibly flexible it is, but also how conceptually simple it is, and this can be seen most clearly with a class diagram:

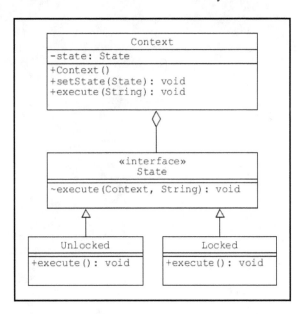

State patterns, like all the patterns in this chapter and other behavioral patterns, are remarkably flexible and this ability to adapted to suit an enormous number of situations is down to their abstract nature. This can make behavioral patterns conceptually more difficult to grasp, but a little trial and error is a good way to find the right patterns for the right situations.

Summary

Behavioral patterns can appear very similar in structure and there is a lot of functional overlap, and this chapter has been largely theoretical so that we could approach them collectively. Once we become familiar with these structures, we will find ourselves returning to them frequently and in many situations.

In the next chapter, we will concentrate on more technical matters and see how to develop apps for the wide variety of form factors available, such as wrist watches and TV screens. From the work we have done so far, we can see how patterns such as the visitor could be used to manage such choices. As we have already experienced, the system manages much of this for us, often using inbuilt patterns of its own. Nevertheless, there are many opportunities to simplify and rationalize our code with design patterns.

11

Wearable Patterns

So far in this book, all the Android applications we have considered have been designed for mobile devices such as phones and tablets. As we have seen, the framework provides great facility for ensuring our designs work well on the wide variety of screen sizes and shapes available. However, there are three form factors that the work we have done so far does not cover, and that is wearable devices such as wrist watches, in-car consoles, and television sets.

When it comes to the application of design patterns to these alternative platforms, which patterns we choose depends on the purpose of the application rather than the platform itself. As we concentrated heavily on patterns in the previous chapter, this chapter will mostly cover the practicalities of building apps for each of these device types. However, as we shall see when we take a look at TV apps, these employ a **model-view-presenter pattern**.

As we have not yet dealt with coding sensors, the chapter will include exploring how a user's heart rate can be read and responded to by our code. The way that physical sensors, such as heart rate monitors and accelerometers, are managed is very similar and by examining one, we can learn how the others are handled.

In this chapter, you will learn how to:

- Set up a TV app
- Use the leanback library
- Apply MVP patterns
- Create banners and media components
- Understand the browser and consumption views
- Connect to wearable devices
- Manage wearable screen shapes
- Handle wearable notifications
- Read sensor data
- Understand Auto safety features
- Configure Auto apps for media services
- Configure Auto apps for messaging services

The first thing to consider when developing for this wide range of form factors is not just the size of the graphics we need to prepare, but also the distance it is viewed from. Most Android devices are used from just a few inches away and are often designed to be rotated, moved, and touched. The exception here is the television screen, which is generally viewed from around 10 feet away.

Android TV

Televisions are generally best for relaxing activities such as watching movies and shows and playing games. However, there is still a large area of overlap, particularly concerning games, where many apps can be easily converted to work on TV. The viewing distance, high definition, and controller devices mean that a few adaptions need to be made and this is largely helped by the leanback support library. This library facilitates the model-view-presenter design pattern, which is an adaption of the model-view-controller pattern.

There is no limit to the types of app that could be developed for TV, but a large percentage of them fall into two categories, games and media. Unlike games, which often benefit from having unique interfaces and controls, media based apps should generally use widgets and interfaces that are familiar and consistent across the platform. This is where the **leanback library** comes in, providing a variety of detail, browser and search widgets, and overlays.

The leanback library is not the only support library that is of use to TV development, both the CardView and RecyclerView are useful and the RecyclerView is in fact required, as some leanback classes depend on it.

Android Studio provides a very useful TV module template that provides a dozen or so classes that demonstrate most of the features needed in many media based TV applications. It is very worthwhile to take a good look at this template as it serves as a rather good tutorial. However, it is not necessarily the best starting point for individual projects unless they are quite generic in nature. If you are planning any original projects, it is necessary to know one or two things about how TV projects are set up, starting with the device home screen.

TV home screen

The home screen is the entry point for Android TV users. From here they can search for content, adjust settings, and access applications and games. The first view the user will get of our app will be on this screen in the form of a banner image.

Every TV app has a banner image. This is a 320 x 180 dp bitmap that should portray what our app does in a simple and efficient manner. For example:

Banners can contain colorful photographic imagery too, but text should always be kept bold and to a minimum. The banner can then be declared in the project manifest. To see how this is done, and how other **manifest** properties relevant to TV apps can be set, follow these steps:

1. Start a new project, selecting **TV** as the **Target Android Device** and **Android TV Activity** as the activity template.
2. Add your image to the drawable folder and call it `banner` or something like that.
3. Open the `manifests/AndroidManifest.xml` file.
4. Delete the following line:

```
android:banner="@drawable/app_icon_your_company"
```

5. Edit the opening `<application>` node to include the following highlighted line:

```
<application
    android:allowBackup="true"
    android:banner="@drawable/banner"
    android:label="@string/app_name"
    android:supportsRtl="true"
    android:theme="@style/Theme.Leanback">
```

6. In the root `<manifest>` node, add the following attribute:

```
<uses-feature
    android:name="android.hardware.microphone"
    android:required="false" />
```

This last `<uses-feature>` node is not strictly required, but will make your app available to older televisions that do not have microphones included. If your app relies on voice control, then omit this attribute.

We will also need to declare a leanback launcher for our main activity, which is done like this:

```
<intent-filter>
  <action
        android:name="android.intent.action.MAIN" />
  <category
        android:name="android.intent.category.LEANBACK_LAUNCHER" />
</intent-filter>
```

If you are building for TV alone, then this is all you need to do in terms of making your app available in the TV section of the Play store. However, you may be developing an application such as a game that can be played on other devices. In this case, also include the following clause to make it available to devices that can be rotated:

```
<uses-feature
    android:name="android.hardware.screen.portrait"
    android:required="false" />
```

In these situations, you should also set `android.software.leanback` to `required="false"` and revert to the material or *appcompat* themes.

You may be wondering why we moved the banner declaration from the main activity to the application as a whole. This was not strictly necessary, and what we have done is simply apply one banner to the whole app, regardless of how many activities it contains. Unless you want separate banners for each activity, this is usually the best way to go.

TV model-view-presenter pattern

The leanback library is one of the few that directly facilitate the use of a design pattern, the model-view-presenter (MVP) pattern, which is a derivation of model-view-controller (MVC). Both these patterns are remarkably simple and obvious, and some might say that they do not really qualify as patterns at all. Even if you had never come across design patterns before, you would have applied one or both of these *architectures*.

We covered MVC and MVP briefly earlier, but to recap, in an MVC pattern the view and the controller are separate. For example, when the controller receives input from the user, such as a click of a button, it passed this to the model which executes its logic and forwards this updated information to the view, which then displays this change to the user, and so on.

The MVP pattern combines the functions of both view and controller, making it an intermediary between the user and the model. This is something we have seen before in the shape of the adapter pattern, in particular the way recycler views and their adapters work.

The leanback presenter class also works in conjunction with a nested view holder, and in terms of the MVP pattern, the view is any Android view and the model can be any Java object or collection of objects we choose. This means that we can use the presenter to act as an adapter between any logic of our choice and any layout we wish.

Despite the freedom of this system, before embarking on project development, it is worth taking a little look at some of the conventions used in TV app development.

TV app structure

A large number of media TV apps offer a limited set of functions, and this is usually all that is required. For the most part, users want to:

- Browse for content
- Search for content
- Consume content

The leanback library provides fragment classes for each of these. A typical **browser view** is provided by the BrowserFragment and the template demonstrates this with a simple example, along with a SearchFragment:

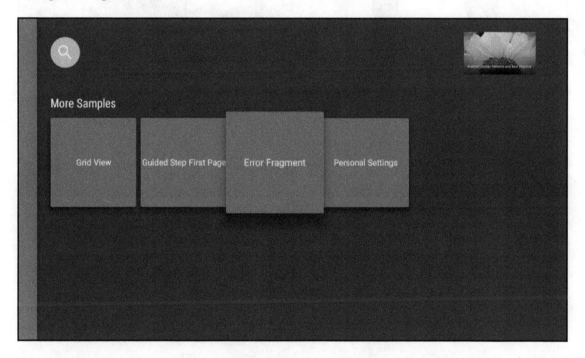

The **consumption view** is provided by the PlaybackOverlayFragment and is probably the simplest of views, comprising little more than a VideoView and the controls.

There is also the `DetailsFragment` that provides content specific information. The content and layout of this view are dependent on the subject matter and can take any form you choose, the regular rules of material design applying. The **design view** scrolls up from the bottom of the consumption view:

The leanback library makes light work of bringing material design to TV devices. If you decide to use views from elsewhere, then the same material rules that apply to them elsewhere apply here too. Before moving on, it is worth mentioning that background images need to have a 5% bleed around the edges to ensure they reach the sides of all TV screens. This means that a 1280 x 720 px image needs to be 1408 x 792 px.

Earlier, we covered the banner image used to launch an app, but we also need a way to direct users to individual content and in particular to familiar or pertinent content.

Recommendation cards

The top row of the Android TV home screen is the **recommendation row**. This allows user to quickly access content based on their viewing history. Content can be recommended because it is a continuation of previously viewed content or related in some way based on the user's viewing history.

When designing recommendation cards, there are only a handful of design factors we need to consider. These cards are constructed from an image or large icon, a title, a subtitle and an application icon, like so:

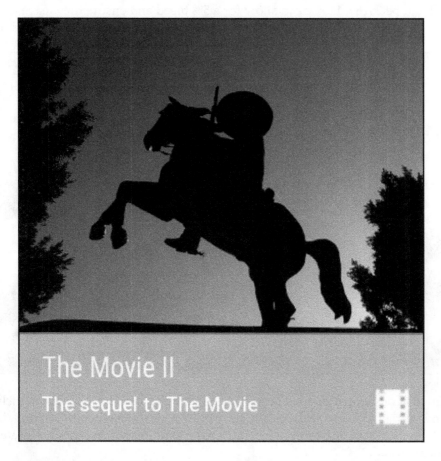

The Movie II

The sequel to The Movie

There is a certain amount of flexibility when it comes to the aspect ratio of the card image. The width of the card must never be less than 2/3 of its height or more than 3/2. There must be no transparent elements within the image and it must not be less than 176 dp in height.

Large expanses of white can be quite harsh on many televisions. If you need large areas of white, use #EEE rather than #FFF.

If you take a look at the recommendation row on a live Android TV set, you will see that as each card is highlighted, the background image changes and we too should provide background images for each recommendation card. These images must differ from the one on the card and be 2016 x 1134 px to allow for a 5% bleed and ensure they leave no gaps around the edge of the screen. These images too should have no transparent sections.

The challenge of designing for such large screens affords us the opportunity to include colorful and vibrant imagery with high-quality graphics. At the other end of this size spectrum falls the wearable device, where space is premium and an entirely different approach is required.

Android Wear

Wearable Android apps deserve special treatment for yet another reason, and that is that nearly all Android Wear applications act as a companion app and work in conjunction with a main module that runs on the user's handset. This tethering is an interesting and straightforward process and many mobile apps can be enhanced greatly by adding a wearable component. Another feature that makes developing for wearables a lot of fun is the availability of exciting new sensors and gadgetry. In particular, the heart rate monitor found in many smart watches has proved unsurprisingly popular in fitness apps.

Wearables are one of the most exciting areas of smart device development. Smart phones and other worn devices with a whole range of new sensors open up uncountable new possibilities for developers.

Apps running on wearable devices need to be connected to a parent application running on a mobile handset and are best thought of as extensions of the main app. Whereas most developers have access to at least one handset, wearable devices can be an expensive option for testing only, particularly because we would need a minimum of two. This is because of the difference in the way **square and round screens** are handled. Fortunately, we can create AVDs with an emulator and connect these to either a real phone or tablet or a virtual one.

Pairing with a wearable device

To best see this difference between round and square screen management, begin by creating an emulator for each:

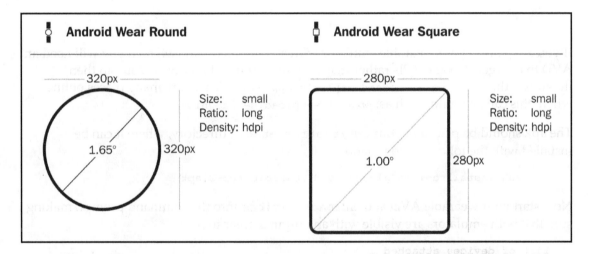

There is also a chinned version, but for programming purposes we can consider this the same as the round screen.

How you pair the wearable AVD will depend on whether you are coupling it with a real handset or another emulator. If you are using a handset, you will need to download the Android Wear app from:

`https://play.google.com/store/apps/details?id=com.google.android.wearable.ap`
`p`

Then locate the `adb.exe` file, which, by default, is located in
`user\AppData\Local\Android\sdk\platform-tools\`

Open the command window here and issue the following command:

adb -d forward tcp:5601 tcp:5601

You can now launch the companion app and follow the instructions to pair the devices.

 You will need to issue this port forwarding command each time you connect the handset.

If you are pairing your wearable emulator with an emulated handset, then you will need an AVD that targets Google APIs rather than a regular Android platform. You can then download the `com.google.android.wearable.app-2.apk`. There are many places online where this can be found, such as: `www.file-upload.net/download`

The apk should be placed in your `sdk/platform-tools` directory, where it can be installed with the following command:

adb install com.google.android.wearable.app-2.apk

Now start your wearable AVD and enter `adb devices` into the command prompt, making sure that both emulators are visible with an output similar to this:

```
List of devices attached
emulator-5554    device
emulator-5555    device
```

Enter:

adb telnet localhost 5554

at the command prompt, where `5554` is the phone emulator. Next, enter `adb redir add tcp:5601:5601`. You can now use the Wear app on the handheld AVD to connect to the watch.

When creating Wear projects, you will need to include two modules, one for the wearable component and one for the handset.

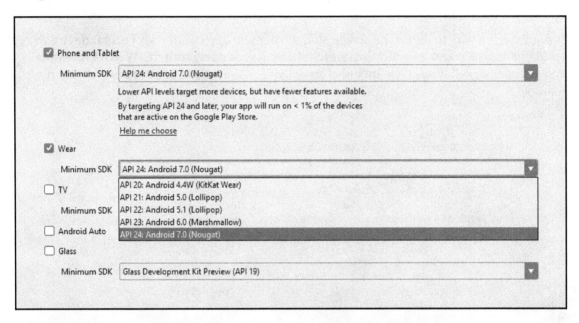

Android provides a **wearable UI support library** that provides some very useful features for Wear developers and designers. If you have created a wearable project using the wizard, this will have been included during setup. Otherwise you will need to include the following dependencies in the `Module: wearbuild.gradle` file:

```
compile 'com.google.android.support:wearable:2.0.0-alpha3'
compile 'com.google.android.gms:play-services-wearable:9.6.1'
```

You will also require these lines in the Module: mobile build file:

```
wearApp project(':wear')
compile 'com.google.android.gms:play-services:9.6.1'
```

Managing screen shapes

We have no idea in advance which of these shapes our apps will be running on, and there are two solutions to this conundrum. The first, and most obvious, is simply to create a layout for each shape and this is very often the best solution. If you have a wearable project created with the wizard, you will see that template activities for both shapes have been included.

We still need a way to detect the screen shape when the app is run on an actual device or emulator so that it knows which layout to inflate. This is done with the **WatchViewStub** and the code to call it has to be included in onCreate() method of our main activity file, like so:

```
@Override
protected void onCreate(Bundle savedInstanceState) {
    super.onCreate(savedInstanceState);
    setContentView(R.layout.activity_main);

    final WatchViewStub stub = (WatchViewStub)
            findViewById(R.id.watch_view_stub);
    stub.setOnLayoutInflatedListener(
            new WatchViewStub.OnLayoutInflatedListener() {

        @Override
        public void onLayoutInflated(WatchViewStub stub) {
            mTextView = (TextView) stub.findViewById(R.id.text);
        }

    });
}
```

This can then be implemented in XML like so:

```
<android.support.wearable.view.WatchViewStub
xmlns:android="http://schemas.android.com/apk/res/android"
    xmlns:app="http://schemas.android.com/apk/res-auto"
    xmlns:tools="http://schemas.android.com/tools"
    android:id="@+id/watch_view_stub"
    android:layout_width="match_parent"
    android:layout_height="match_parent"
    app:rectLayout="@layout/rect_activity_main"
    app:roundLayout="@layout/round_activity_main"
    tools:context=".MainActivity"
    tools:deviceIds="wear">
  </android.support.wearable.view.WatchViewStub>
```

The alternative to creating separate layouts for each screen shape is to use a layout that itself is aware of screen shape. This comes in the form of the **BoxInsetLayout**, which adjusts padding settings for round screens and only positions views within the largest possible square within that circle.

The BoxInsetLayout can be used like any other layout, as the root ViewGroup in your main XML activity:

```
<android.support.wearable.view.BoxInsetLayout
    xmlns:android="http://schemas.android.com/apk/res/android"
    xmlns:app="http://schemas.android.com/apk/res-auto"
    android:layout_height="match_parent"
    android:layout_width="match_parent">

    . . .

</android.support.wearable.view.BoxInsetLayout>
```

There are definitely drawbacks to this approach, as it does not always make the most of the space available on round faces, but what the BoxInsetLayout lacks in flexibility it makes up for in ease of use. In most cases, this is not a drawback at all, as a well-designed Wear app should only grab the user's attention briefly with simple information. Users are not keen to navigate complex UIs on their watches. The information we display on a watch screen should be able to be absorbed in a glance and the responding action should be limited to no more than a tap or a swipe.

One of the main uses of smart devices is to receive notifications when the user is otherwise unable to access their handset, for example when they are exercising.

Wearable notifications

There is very little to adding wearable notification functionality to any mobile app. Recall how notifications are delivered from Chapter 9, *Observing Patterns*:

```
private void sendNotification(String message) {

    NotificationCompat.Builder builder =
            (NotificationCompat.Builder)
            new NotificationCompat.Builder(this)
                    .setSmallIcon(R.drawable.ic_stat_bun)
                    .setContentTitle("Sandwich Factory")
                    .setContentText(message);

    NotificationManager manager =
            (NotificationManager)
```

```
                  getSystemService(NOTIFICATION_SERVICE);
        manager.notify(notificationId, builder.build());

        notificationId += 1;
    }
```

To adapt this to also send the notification to the paired wearable device, simply add these two lines to the builder string:

```
.extend(new NotificationCompat.WearableExtender()

.setHintShowBackgroundOnly(true))
```

The optional `setHintShowBackgroundOnly` setting allows us to display the notification without a background card.

Most of the time a wearable is used as an output device, but it can also act as an input and many new functions can be derived when sensors are placed close to the body, such as the heart rate monitor included in many smart phones.

Reading sensors

There are a growing number sensors available on most smart devices today and smart watches offer new opportunities to developers. Fortunately these sensors are very simple to program, they are after all, just another input device and as such, we employ listeners to *observe* them.

Although the function and purpose of individual sensors differs widely, the way they are read is almost identical, the only difference being the nature of their outputs. Here we will look at the heart rate monitor found on many wearables:

1. Open or start a Wear project.
2. Open the wear module and add a BoxInsetLayout with a TextView to the main activity XML file, like so:

```
<android.support.wearable.view.BoxInsetLayout
    xmlns:android="http://schemas.android.com/apk/res/android"
    xmlns:app="http://schemas.android.com/apk/res-auto"
    android:layout_height="match_parent"
    android:layout_width="match_parent">

    <TextView
        android:id="@+id/text_view"
        android:layout_width="match_parent"
```

```
            android:layout_height="wrap_content"
            android:layout_gravity="center_vertical" />
      </android.support.wearable.view.BoxInsetLayout>
```

3. Open the Manifest file in the wear module and add the following permission inside the root `manifest` node.

```
<uses-permission android:name="android.permission.BODY_SENSORS" />
```

4. Open the main Java activity file in the wear module and add the following fields:

```
private TextView textView;
private SensorManager sensorManager;
private Sensor sensor;
```

5. Implement a `SensorEventListener` on the activity:

```
public class MainActivity extends Activity
            implements SensorEventListener {
```

6. Implement the two methods required by the listener.
7. Edit the `onCreate()` method like this:

```
@Override
protected void onCreate(Bundle savedInstanceState) {
    super.onCreate(savedInstanceState);
    setContentView(R.layout.activity_main);

    textView = (TextView) findViewById(R.id.text_view);

    sensorManager = ((SensorManager)
            getSystemService(SENSOR_SERVICE));
    sensor = sensorManager.getDefaultSensor
            (Sensor.TYPE_HEART_RATE);
}
```

8. Add this `onResume()` method:

```
protected void onResume() {
    super.onResume();

    sensorManager.registerListener(this, this.sensor, 3);
}
```

9. And this `onPause()` method:

```
@Override
protected void onPause() {
    super.onPause();

    sensorManager.unregisterListener(this);
}
```

10. Edit the `onSensorChanged()` callback, like so:

```
@Override
public void onSensorChanged(SensorEvent event) {
    textView.setText(event.values[0]) + "bpm";
}
```

As you can see, sensor listeners act like observers in exactly the same way as click and touch listeners. The only real difference is that sensors need to be explicitly registered and unregistered, as they are not available by default and need to be switched off when done with to preserve battery.

All sensors can be managed the same way with a sensor event listener and it is usually best to check for each sensor's presence when initializing the app with:

```
private SensorManager sensorManagerr = (SensorManager)
getSystemService(Context.SENSOR_SERVICE);
    if (mSensorManager.getDefaultSensor(Sensor.TYPE_ACCELEROMETER) !=
null){

    . . .
    }
    else {
    . . .

}
```

Wearable devices open up a whole new world of app possibilities, bringing Android into ever increasing aspects of our lives. Another such example would be the use of Android devices in our cars.

Android Auto

As with Android TV, Android Auto can run many apps designed originally for mobile devices. Of course, with in-car software, safety is the overwhelming priority, which is why most Auto apps concentrate on audio functions, such as messaging and music.

 Because of the emphasis on safety, Android Auto apps have to undergo stringent testing before they can be published.

It barely needs mentioning that safety is the overriding principle when developing in-car apps, and for this reason, Android Auto applications nearly all fall into two categories: music or audio players and messaging.

All applications require extensive testing during the development phase. Obviously, it would be impractical and highly dangerous to test an Auto app on a live device, and for this reason Auto API simulators are provided. These can be installed from the SDK manager's tools tab.

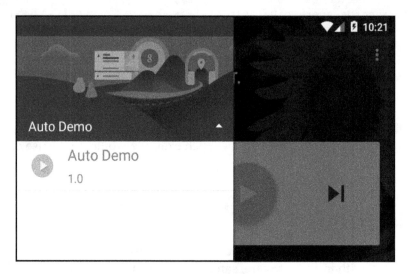

Auto safety considerations

Many of the rules governing Auto safety are simple common sense, avoid animations, distractions and delays, but of course it is necessary to formalize these and Google have done so. These rules concern driver attention, screen layout, and readability. The most significant can be found in the list here:

- There must be no animated elements on the Auto screen
- Only audio ads are allowed
- Apps must support voice control
- All buttons and clickable controls must respond within two seconds
- Text must be longer than 120 characters and must always be in the default Roboto font
- Icons must be white so that the system can control contrast
- Apps must support day and night mode
- App must support voice commands
- App-specific buttons must respond to user actions with no more than a two-second delay

A comprehensive list can be found at:

`developer.android.com/distribute/essentials/quality/auto.html`

IMPORTANT: These, and several other stipulations, will be tested by Google before publication, so it is essential that you run all of these tests yourself.

 Designing apps that are suitable for day and night modes and that can have contrast controlled by the system to automatically remain readable in different light conditions is quite a detailed subject and Google have produced a very useful guide to this, which can be found at: `commondatastorage.googleapis.com/androiddevelopers/shareables/au to/AndroidAuto-custom-colors.pdf`

Other than safety and the limitation of application type, Auto apps differ from the other apps we have explored only in how they are set up and configured.

Configuring Auto apps

If you use the studio wizard to set up an Auto app, you will see that, as with Wear apps, we have to include both a Mobile and an Auto module. Unlike wearable projects, this does not involve a second module and everything is managed from the mobile module. The addition of the Auto component provides a configuration file that can be found in `res/xml`. For example:

```
<?xml version="1.0" encoding="utf-8"?>
<automotiveApp>
    <uses name="media" />
</automotiveApp>
```

For messaging apps, we would use:

```
    <uses name="media" />
```

The other important Auto elements can be found by examining the template-generated manifest files. Whichever type of app you choose to develop, you will need to add the following metadata:

```
<meta-data
    android:name="com.google.android.gms.car.application"
    android:resource="@xml/automotive_app_desc" />
```

As you would imagine, a music or audio provider requires a service alongside the launcher activity and a messaging app would require a receiver. A music service tag would look like the following snippet:

```
<service
    android:name=".SomeAudioService"
    android:exported="true">
    <intent-filter>
        <action android:name="android.media.browse.MediaBrowserService" />
    </intent-filter>
</service>
```

For a messaging application, we need a service and two receivers, one to receive messages and one to send them, like so:

```
<service android:name=".MessageService">
</service>

<receiver android:name=".MessageRead">
    <intent-filter>
        <action android:name="com.kyle.someapplication.ACTION_MESSAGE_READ"
/>
    </intent-filter>
</receiver>

<receiver android:name=".MessageReply">
    <intent-filter>
        <action
android:name="com.kyle.someapplication.ACTION_MESSAGE_REPLY" />
    </intent-filter>
</receiver>
```

In-car devices represent one of the largest growth areas of Android development and this is set to grow further still as hands-free driving becomes more prevalent. Very often, we may only want to include a single Auto feature into an app designed mostly for other form factors.

Unlike handheld and wearable devices, we do not have to concern ourselves overly with screen size, shape or density, nor do we have to worry about the make or model of a particular vehicle. This will no doubt change in the near future as the nature of driving and transportation changes.

Summary

The alternative form factors described in this chapter provide exciting new platforms for developers and the kind of applications we can create. This is not simply a matter of coming up with apps for each platform, and it is perfectly possible to include all three device types within a single app.

Take the sandwich building app we looked at earlier; we could easily adapt it so that users could order a sandwich when they were watching a movie. Equally, we could send a notification of when their order was ready to their smart phone or auto console. In short, these devices open up the market for new apps and added functionality for existing ones.

However clever or versatile our creations, there are very few apps that couldn't benefit from the promotional opportunities provided my social media. A single *tweet* or *like* can reach untold numbers without the expense of advertising.

In the next chapter, we will see how easy it is to add social media features to our apps and also how we can build web-app functionality into Android apps or even construct complete web apps, using the SDK's webkit and WebView.

12
Social Patterns

So far in this book, we have covered many of the aspects of mobile app development. However, even the best designed and most useful application can benefit enormously from employing social media and other web content.

The sandwich builder app we covered in earlier chapters is a good example of an app that could have its circulation boosted by generating Facebook likes and tweets, and these and other social media all provide technologies to incorporate such features directly into our app.

As well as incorporating existing social media platforms into our apps, we can embed any web content we like right into an activity with the **WebView** class. This extension of the view class can be used to add a single web page to an app or even build a complete web application. The WebView class is extremely useful when we have products or data that need regular updating, as this can be achieved without having to recode and release updates.

We will start this chapter by taking a look at the WebView class and see how we can incorporate JavaScript to give pages functionality; and then we'll explore some of the social media SDKs that allow us to incorporate many of their features, such as sharing and posting and liking.

In this chapter, you will learn how to do the following:

- Open a web page in a WebView
- Open a web page in a browser
- Enable and use JavaScript
- Use a JavaScriptInterface to bind script with native code
- Write efficient HTML for web apps
- Create a Facebook app
- Add a LikeView button
- Create a Facebook sharing interface
- Integrating Twitter
- Sending tweets

Adding web pages

Including a single web page in an activity or fragment using the WebView class is almost as simple as adding any other kind of view. There are three easy steps, as follows:

1. Add the following permission to the manifest:

```
<uses-permission
    android:name="android.permission.INTERNET" />
```

2. The `WebView` itself looks like this:

```
<WebView xmlns:android="http://schemas.android.com/apk/res/android"
    android:id="@+id/web_view"
    android:layout_width="match_parent"
    android:layout_height="match_parent" />
```

3. Finally, the Java for adding page is as follows:

```
WebView webView = (WebView) findViewById(R.id.web_view);
webView.loadUrl("https://www.packtpub.com/");
```

That's all there is to it, although you would probably want remove or reduce the default 16dp margins for most pages.

This system is ideal for when dealing with pages that have been specifically designed for our app. If want to send our user to any other web page, then it is considered better practice to use a link, so that the user can open it with their chosen browser.

Including a link

To do this, any clickable view can act as the link, and the click listener can then respond like so:

```
@Override
    public void onClick(View v) {
        Intent intent = new Intent();
        intent.setAction(Intent.ACTION_VIEW);
        intent.addCategory(Intent.CATEGORY_BROWSABLE);
        intent.setData(Uri.parse("https://www.packtpub.com/"));
        startActivity(intent);
    }
```

We can see that the correct use of web views is to use pages that are specifically designed to be an integral part of our app. Although the user needs to know that they are online (as charges may apply) our web views should look and behave like the rest of the app. It is perfectly possible to have more than one web view on a screen and to mix them with other widgets and views, and if we are working on an app that stores user details, this is often more easily managed using web tools rather than the Android APIs.

The WebView class comes with a comprehensive array of settings that can be used to control a lot of properties, such as zoom functionality, image loading, and display settings.

Configuring WebSettings and JavaScript

Although we can design our web views to look like other app components, they do of course posses a large number of web-specific properties and can, as web elements, be navigated as one would in a browser. These and other settings are managed elegantly by the **WebSettings** class.

This class consists largely of a long series of setters and getters. The entire collection can be initialized like so:

```
WebView webView = (WebView) findViewById(R.id.web_view);
WebSettings webSettings = webView.getSettings();
```

We can now use this object to query the sate of our web views and to configure them to our wishes. For example, JavaScript is disabled by default but can be changed easily:

```
webSettings.setJavaScriptEnabled(true);
```

There is a large number of such methods, all of which are listed in the documentation:

developer.android.com/reference/android/webkit/WebSettings.html

These settings are not the only way we can take control over our web view, and it has some very useful methods of its own, most of which are listed here:

- `getUrl()` – Returns the web view's current URL

- `getTitle()` – Returns the page's title if specified in the HTML

- `getAllAsync(String)` – Simple search function, highlighting occurrences of the given string

- `clearHistory()` – Empties the current history cache

- `destroy()` – Closes and empties the web view

- `canGoForward()` and `canGoBack()` – Enables the local history stack

These methods, along with web settings, allow us to do a lot more with a web view than simply access changeable data. We can, with a little effort, provide much of the functionality of a web browser.

Whether we choose to present our web views as a seamless part of our app or provide the user with a fuller internet based experience, we will most likely want to include some JavaScript in our pages. We saw earlier how to enable JavaScript, but this only allows us to run standalone script; what would be far better would be if we could call an Android method from JavaScript, and this is exactly what the `JavaScriptInterface` does.

The use of an interface like this, to manage the natural incompatibilities between the two languages, and is of course a classic example of the **adapter design pattern**. To see how this can be achieved, follow these steps:

1. Add the following fields to whichever activity you are using for the task:

```
public class WebViewActivity extends Activity {
    WebView webView;
    JavaScriptInterface jsAdapter;
```

2. Edit the `onCreate()` method like so:

```
@Override
public void onCreate(Bundle savedInstanceState) {
    super.onCreate(savedInstanceState);
    setContentView(R.layout.main);

    webView = (WebView) findViewById(R.id.web_view);

    WebSettings settings = webView.getSettings();
    settings.setJavaScriptEnabled(true);

    jsAdapter = new JavaScriptInterface(this);
    webView.addJavascriptInterface(jsAdapter, "jsAdapter");

    webView.loadUrl("http://someApp.com/somePage.html");
}
```

3. Create the adapter class (this can also be an inner class). The `newActivity()` method could be anything we chose. Here, just by way of example, it starts a new activity:

```
public class JavaScriptInterface {
    Context context;

    JavaScriptInterface(Context c) {
        context = c;
    }

    // App targets API 16 and higher
    @JavascriptInterface
    public void newActivity() {
        Intent i = new Intent(WebViewActivity.this,
            someActivity.class);
        startActivity(i);
    }
}
```

4. All that remains is to write the JavaScript to call our native method. Any clickable HTML object will do here. Create the following button on your page:

```
<input type="button"
    value="OK"
    onclick="callandroid()" />
```

5. Now, just define the function in your script, like so:

```
<script type="text/javascript">

    function callandroid() {
        isAdapter.newActivity();
    }

</script>
```

This process is wonderfully easy to implement and makes the web view a very powerful component, and the ability to call our Java methods from our web pages means we can combine web functionality to any app without having to compromise mobile functionality.

Although you will require no assistance in building web pages, there are one or two points that need making regarding best practice.

Writing HTML for WebViews

It is tempting to think that the design of a mobile web app would follows similar conventions to the design of mobile web pages, and in many ways it does, but there are one or two subtle differences that the following list points out:

- Make sure that you are using the correct DOCTYPE, which in our case is this:

```
<?xml version="1.0" encoding="UTF-8"?>
<!DOCTYPE html PUBLIC "-//W3C//DTD XHTML Basic 1.1//EN"
    "http://www.w3.org/TR/xhtml-basic/xhtml-basic11.dtd">
```

- Creating separate CSS and script files can cause a slow connection. Keep this code inline, ideally inside the head or at the very end of the body. Sadly, this means that we have to avoid CSS and web frameworks, and features such as material design have to be coded manually.
- Avoid horizontal scrolling where possible. If your app absolutely requires this then use tabs, or better still, a sliding navigation drawer.

As we have seen, the WebView is a powerful component and makes sophisticated mobile/web hybrid apps very easy to develop. The subject is a large one and one could realistically devote an entire book to the subject. For now, though, it is enough just to understand the scope and power of this tool.

Using built-in web tools is just one way that we can harness the power of the Internet. Being able to connect to social media is probably the most efficient and cheapest methods of promoting a product. One of the most useful and simplest to set up is Facebook.

Connecting with Facebook

Not only is Facebook one of the largest social networks, it is also very nicely set up to assist those wishing to promote their products. The ways that this can work varies from providing automatic logins, customizable advertising, and the ability for users to share products they *like* with others.

To incorporate Facebook features into our Android applications, we will need the **Facebook SDK for Android**, and to make the most of this we will also need a Facebook App ID, which will require creating a simple app on Facebook:

Adding the Facebook SDK

The first step in adding Facebook functionality to our apps is to download the Facebook SDK. This can be found here:

```
developers.facebook.com/docs/android
```

The SDK is a powerful suite of tools, including views, classes, and interfaces that Android developers will be very familiar with. The Facebook SDK can be thought of as a useful extension of our native SDK.

A handy quick start guide can be found on the Facebook developer pages, but as is always the case in such situations, it is far more instructive to follow the process manually, as the following steps demonstrate:

1. Start a new Android Studio project with a minimum API level of 15 or higher.
2. Open the modular `build.gradle` file and make the changes highlighted here:

```
repositories {
    mavenCentral()
}

dependencies {

    . . .

    compile
        'com.android.support:appcompat-v7:24.2.1'
    compile
        'com.facebook.android:facebook-android-sdk:(4,5)'
    testCompile 'junit:junit:4.12'
}
```

3. Add the following permission to the manifest file:

```
<uses-permission
    android:name="android.permission.INTERNET" />
```

4. Then import the following libraries to your main activity or application class:

```
import com.facebook.FacebookSdk;
import com.facebook.appevents.AppEventsLogger;
```

5. Finally, initialize the SDK from the `onCreate()` method of your launch activity, like so:

```
FacebookSdk.sdkInitialize(getApplicationContext());
AppEventsLogger.activateApp(this);
```

This is not all that we need to progress, but before we can go any further, we will need a Facebook App ID, which we can only acquire by creating an app on Facebook.

Obtaining a Facebook App ID

As you will have seen, Facebook apps can be very sophisticated, and their functions are limited only by their creator's imagination and coding prowess. They can, and often are, nothing more than a simple page, and when our emphasis is on an Android app, we need only the simplest of Facebook apps.

For now, use the Facebook quick start process, which can be found here:

`https://developers.facebook.com/quickstarts`

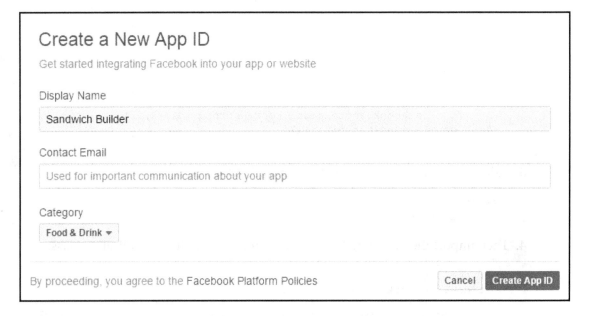

Once you click on **Create App ID**, you will be taken to your developer dashboard. The App ID can be found in the top-left of the window. The following two steps demonstrate how to complete the process we started earlier:

1. Open the `res/values/strings.xml` file and add the following value:

```
<string
    name="facebook_app_id">APP ID HERE</string>
```

2. Now add the following meta-data to the application tag of the manifest:

```
<meta-data
    android:name="com.facebook.sdk.ApplicationId"
    android:value="@string/facebook_app_id" />
```

This completes the process of connecting our Android app to its Facebook counterpart, but we need to compliment this connection by providing information about our mobile app to our Facebook app.

To do this, we will need to return to our Facebook developer dashboard and select **Developer Settings** from your profile (top right) drop-down and then the **Sample App** tab. This will request you enter your package name, launching activity, and **hash key**.

If you are developing an app you intend to publish or use the same hash key for all your projects, you will know it, or have it to hand. Otherwise, the following code will find it for you:

```
PackageInfo packageInfo;

packageInfo = getPackageManager()
        .getPackageInfo("your.package.name",
        PackageManager.GET_SIGNATURES);

for (Signature signature : packageInfo.signatures) {

    MessageDigest digest;
    digest = MessageDigest.getInstance("SHA");
    digest.update(signature.toByteArray());
    String key = new
            String(Base64.encode(digest.digest(), 0));

    System.out.println("HASH KEY", key);
}
```

If you enter this code directly, Studio will offer you a selection of which libraries to import via the quick-fix facility. The correct choices are as follows:

```
import android.content.pm.PackageInfo;
import android.content.pm.PackageManager;
import android.content.pm.Signature;
import android.util.Base64;

import com.facebook.FacebookSdk;
import com.facebook.appevents.AppEventsLogger;

import java.security.MessageDigest;
```

There is more to it than one might imagine, but our app is now connected to Facebook, and we can take advantage of all the promotional opportunities. One of the most important of these is the Facebook Like button.

Adding a LikeView

As you would imagine, the Facebook SDK comes equipped with the traditional *like* button. This is provided as a view and can be added like any other view:

```
<com.facebook.share.widget.LikeView
        android:id="@+id/like_view"
        android:layout_width="wrap_content"
        android:layout_height="wrap_content"/>
```

As with other views and widgets, we can then modify this view from within a Java activity. There are quite a lot of things we can do with this and other Facebook views, and Facebook documents these thoroughly. The LikeView documentation, for example, can be found here:

developers.facebook.com/docs/reference/android/current/class/LikeView

For now, we can just consider what it is the user is liking. This achieved with the `setObjectId()` method, which takes a string argument that can be either your app ID or a URL, like this:

```
LikeView likeView = (LikeView) findViewById(R.id.like_view);
likeView.setObjectId("Facebook ID or URL");
```

There are one or two differences between in-app Like views and those found on the Web. Unlike web likes, the Android like view will not inform the user how many other users have also clicked like and on a device that does not have Facebook installed, our like view will not work at all. These limitations of the Android LikeView are easily countered by using a WebView to contain the like view, which will then act like it would on the web.

The LikeView gives us and users the opportunity to see how popular a particular item is, but to really harness the power of this social platform, we want users to promote us by the modern version of word-of-mouth, that is, by *sharing* our products with their friends.

Content builders

Having a large number of likes is a great way to drive traffic your way, but there is an economy of scale at work here that favors apps with very large numbers of downloads. Apps do not have to be huge to be successful, especially if they are providing a personal or local service, such as delivering custom-made sandwiches. In these cases, a label stating that only 12 people *like* something is not much of a recommendation. However, if those same people share how great their sandwich is with their friends, then we have a very powerful advertising tool at our disposal.

One of the main things that has made Facebook such a successful platform is its understanding that human beings are more interested and influenced by their friends than nameless strangers, and for small to medium enterprises, this can be invaluable. At its simplest we can simply add a share button, just as we did the like button, and this will open the share dialog. The **ShareButton** is as easy to add as the LikeView, as can be seen here:

```
<com.facebook.share.widget.ShareButton
    android:id="@+id/share_button"
    android:layout_width="wrap_content"
    android:layout_height="wrap_content"/>
```

We will also need to set up a content provider in our manifest. The following code should be inserted into the root node:

```
<provider
    android:authorities="com.facebook.app.FacebookContentProvider{
        your App ID here
    }"
        android:name="com.facebook.FacebookContentProvider"
        android:exported="true"/>
```

Unlike the like view, with sharing we have more choice over the content type that we share and we can choose between sharing links, images, videos, and even multimedia.

The Facebook SDK provides a class for each content type and a builder for combining more than one item into a single shareable object.

When sharing photos or images, the `SharePhotoContent` class uses the Bitmap object, which is a more sophisticated and parcelable image format than the drawables we have been using so far. Although there are many ways to create a bitmap, including dynamically from code, it is also relatively simple to convert any of our drawables into a bitmap, as demonstrated in this snippet:

```
Context context;
Bitmap bitmap;
bitmap = BitmapFactory.decodeResource(context.getResources(),
        R.drawable.some_drawable);
```

This can then be defined as shareable content in these two simple steps:

```
// Define photo to be used
SharePhoto photo = new SharePhoto.Builder()
        .setBitmap(bitmap)
        .build();

// Add one or more photos to the shareable content
SharePhotoContent content = new SharePhotoContent.Builder()
        .addPhoto(photo)
        .build();
```

The `ShareVideo` and `ShareVideoContent` classes work in an almost identical fashion and use the file's URI as its source. If you have not worked with video files and URIs before, the simplest way to include them is explained in these short steps:

1. If you have not done so already, create a folder called `raw` directly inside your `res` directory.
2. Place your video(s) in this folder.
3. Make sure the filename contains no spaces or capital letters and is an accepted format, such as `mp4`, `wmv`, or `3gp`.
4. The following code can then be used to extract the video's URI:

```
VideoView videoView = (VideoView)context
        .findViewById(R.id.videoView)
String uri = "android.resource://"
        + getPackageName()
        + "/"
        + R.raw.your_video_file;
```

5. This URI can now be used to define our shared video content, like so:

```
ShareVideo = new ShareVideo.Builder()
        .setLocalUrl(url)
        .build();

ShareVideoContent content = new ShareVideoContent.Builder()
        .setVideo(video)
        .build();
```

These techniques are very handy for sharing single items, and even several items of the same kind, but there are of course times when we would like to mix content, and this can be achieved with the more generic Facebook SDK `ShareContent` class. The following code demonstrates how this can be done:

```
// Define photo content
SharePhoto photo = new SharePhoto.Builder()
    .setBitmap(bitmap)
    .build();

// Define video content
ShareVideo video = new ShareVideo.Builder()
    .setLocalUrl(uri)
    .build();

// Combine and build mixed content
ShareContent content = new ShareMediaContent.Builder()
```

```
        .addMedium(photo)
        .addMedium(video)
        .build();

ShareDialog dialog = new ShareDialog(...);
dialog.show(content, Mode.AUTOMATIC);
```

These simple classes provide a flexible way to allow users to share content with their friends. There is also a send button that allows users to share our content privately with individuals or groups, and although useful to the user, this function serves little commercial purpose.

A valuable tool when testing shared content is provided by the Facebook Sharing Debugger, which can be found here:

```
developers.facebook.com/tools/debug/sharing/?q=https%3A%2F%2Fdevelopers.fac
ebook.com%2Fdocs%2Fsharing%2Fandroid
```

This is particularly useful, as there is no other simple way to see how our shared content is actually viewed by others.

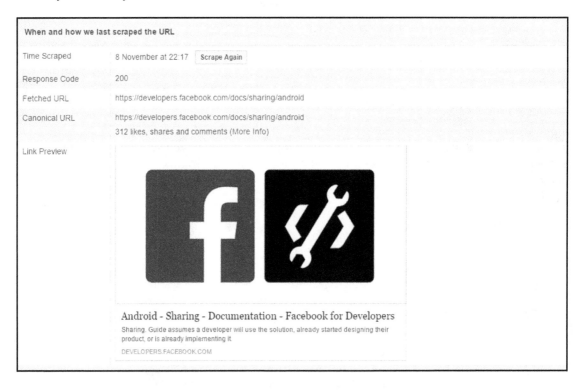

Facebook is not only one of the most popular social networks; it is also has a very well-thought-out SDK and is probably the most developer-friendly social network. This, of course, is no reason to ignore the others, chief among which is Twitter.

Integrating Twitter

Twitter provides a very different social platform to Facebook, and people use it very differently as well. It is, however, another powerful tool in our arsenal, and like Facebook, it offers unparalleled promotional opportunities.

Twitter employs a powerful framework integration tool called **Fabric** that allows developers to integrate Twitter functionality into our applications. Fabric can be downloaded directly into Android Studio as a plugin. Before downloading the plugin, it is necessary to register with Fabric. This is free and can be found at fabric.io.

Once registered, open Android Studio and then select **Browse Repositories...** from **Settings > Plugins**:

Once installed, Fabric has walk-through tutorial system and requires no further instruction. However it is not necessary to use this framework at all if all your app needs to do is post single tweets as this can be achieved with the vanilla SDK.

Sending tweets

Fabric is a sophisticated tool and, thanks to its inbuilt tuition, has a fast learning curve, but it still takes time to master and provides a lot of functionality that most apps won't need. If all you want to do is have your app post a single tweet, it can be done without Fabric, like this:

```
String tweet
        = "https://twitter.com/intent/tweet?text
        =PUT TEXT HERE &url="
        + "https://www.google.com";
Uri uri = Uri.parse(tweet);
startActivity(new Intent(Intent.ACTION_VIEW, uri));
```

Even if all we ever do with Twitter is send tweets, this is still a very useful social function, and if we choose to take advantage of Fabric, we can build apps that rely heavily on Twitter, posting live streams and performing complex traffic analysis. As with Facebook, it is also always a good idea to think about what can be achieved with a web view, and building partial web apps into our mobile one is often the simplest solution.

Summary

Integrating social media into our mobile apps is a powerful tool and can make all the difference to the success of an app. In this chapter, we have seen how Facebook and Twitter provide software tools to facilitate this, and of course, other social media, such as Instagram and WhatsApp, provide similar developer tools.

Social media is an ever-changing world, and new platforms as well as development tools are appearing all the time, and there is no reason to believe that Twitter and even Facebook might one day go the way of MySpace. This is yet another reason to consider using WebViews where possible: creating simple web apps within our main apps allows us a higher degree of flexibility.

This brings us almost to the end of our journey, and in the next chapter we will look at what is generally the final stage of development, publication. However, this is also the point where we have to consider potential income, particularly advertising and in-app purchasing.

13
Distribution Patterns

With most of the important aspects of Android development covered, we are left with just the processes of deploying and publishing. Simply getting an app published on the Google Play store is not a complex process, but there are a few tips and tricks that we can apply to maximize an app's potential reach, and of course, there is a growing number of ways to make money from our apps.

In this chapter, we will look at how to increase backwards compatibility beyond that provided by the support libraries, and then move on to see how the registration and distribution processes work, and then we will explore the various ways to make our application pay.

In this chapter, you will learn how to do the following:

- Prepare an app for distribution

- Generate a digital certificate

- Register as a Google Developer

- Prepare promotional material

- Publish an app on Google Play store

- Incorporate in-app billing

- Include advertisements

Extending platform scope

The support libraries we have been working with throughout the book do a marvelous job of making our apps available on older devices, but they do not work for all situations, and many new innovations simply cannot be realized on some older machines. Taking a look at the following device dashboard, it is obvious that we would like to extend our apps back to API level 16:

Version	Codename	API	Distribution
2.2	Froyo	8	0.1%
2.3.3 - 2.3.7	Gingerbread	10	1.3%
4.0.3 - 4.0.4	Ice Cream Sandwich	15	1.3%
4.1.x	Jelly Bean	16	4.9%
4.2.x		17	6.8%
4.3		18	2.0%
4.4	KitKat	19	25.2%
5.0	Lollipop	21	11.3%
5.1		22	22.8%
6.0	Marshmallow	23	24.0%
7.0	Nougat	24	0.3%

We have seen how the AppCompat library enables our apps to run on platforms even older than this, but we have to avoid using some features. For example, the `view.setElevation()` method (along with other material features) will not work below API level 21 and will cause the machine to crash if it is called.

It would be tempting to think that we could simply sacrifice such features for the benefit of reaching a wider audience, but fortunately, this is not necessary as it is possible to detect dynamically which platform our app is running on with the following conditional clause:

```
if (Build.VERSION.SDK_INT >= Build.VERSION_CODES.LOLLIPOP) {
    someView.setElevation(someValue);
}
```

It is always down to the individual developer, but this slight drop in quality is often well worth the large increase in potential user adoption.

The preceding example is a simple one, however, and adding this kind of dynamic backwards compatibility can often require a lot of extra coding. A good example might be the camera2 API, which is far more sophisticated than its predecessor but only available on devices carrying API 21 and higher. In such a case, we can apply exactly the same principle but would need to set up a more sophisticated system. The clause might cause different methods to be called or even different activities to be launched.

However, we choose to implement this, we can of course, employ design patterns. There are several that could be used here, but the most suitable would probably be the strategy pattern along the lines of the one seen here:

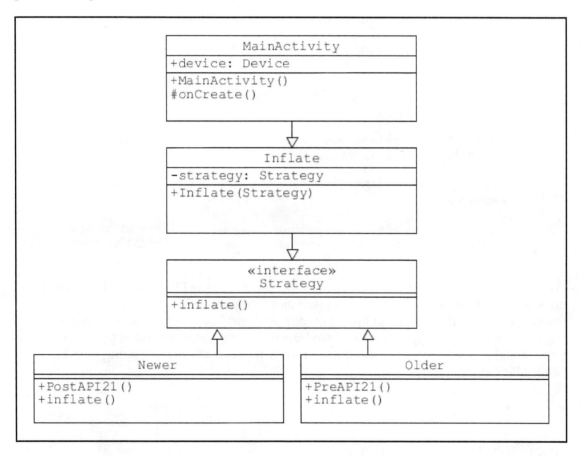

This approach may often require a considerable amount of extra coding, but the expanded potential market often makes that well worth the extra work. Once we have set the scope of our app like this, it is then ready to publish.

Publishing apps

It goes without saying that you will have exhaustively tested your app on a wide variety of handsets and emulators and probably prepared your promotional material and checked out Google Play Policies and Agreements. There are many things to consider before publication, such as content rating and country distribution. From a programming point of view, there are just three things that we need to check before we proceed:

- Remove all logging from the project such as the following:

```
private static final String DEBUG_TAG = "tag";
Log.d(DEBUG_TAG, "some info");
```

- Make sure you have an application label and icon declared in your manifest. Here's an example:

```
android:icon="@mipmap/my_app_icon"
android:label="@string/my_app_name"
```

- Ensure you have declared all the necessary permissions in the manifest. Here's an example:

```
<uses-permission android:name="android.permission.INTERNET" />
<uses-permission android:name="android.permission.ACCESS_NETWORK_STATE" />
```

We are now just three steps from seeing our app on the Google Play store. All we need to do is generate a Signed Release APK, register as a Google Play Developer and, finally upload our app to the store or publish it on our own site. There are also one or two other ways of publishing an app and we will see how they are done at the end of the section. First, though, we will begin by generating an APK that is ready for uploading onto the Google Play store.

Generating a signed APK

All published Android apps require a digitally signed certificate. This is used to prove the authenticity of an app. Unlike many other digital certificates, there is no authority and you hold the signed key, which clearly has to be securely protected. To do this, we need to generate a private key and then use it to generate a signed APK. There are some very handy tools on GitHub for facilitating this process, but here, to aid our understanding, we will follow the traditional route. This can all be done in the Android Studio with the Generate Signed APK Wizard. These steps will take you through it:

1. Open the app you want to publish.
2. Start the Generate Signed APK Wizard from the **Build | Generate Signed APK...** menu.
3. Select **Create new...** on the first screen.
4. On the next screen, provide a path and name for your key store along with a strong password.
5. Do the same for the Alias.
6. Select a Validity of greater than 27 years, like so:

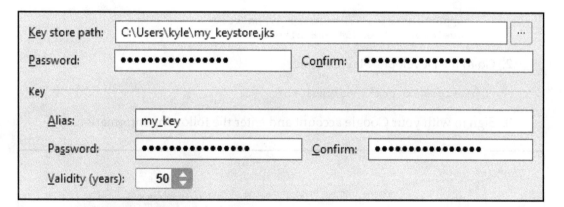

7. Fill in at least one of the Certificate fields. Click on **OK**, and you will be taken back to the wizard.
8. Select **release** as the Build Variant and click on **Finish**.
9. You now have a signed APK ready for publication.

The key store (a `.jks` file) can be used to store any number of keys (aliases). It is perfectly fine to use the same key for all your apps, and you must use the same key when producing updates of an app. Google require certificates to be valid until at least 22 October 2033, and any number that surpasses this date will suffice.

IMPORTANT
Keep at least one secure backup of your keys. If you lose them, you will not be able to develop future versions of those apps.

Once we have our digital signature, we are ready to register as a developer with Google.

Registering as a developer

As with signing an APK, registering as a developer is similarly straightforward. Note that Google charge a one-off fee of USD 25 and 30% of any revenue your app may generate. The following directions assume that you have already have a Google account:

1. Review **Supported Locations** at the following URL:

   ```
   support.google.com/googleplay/android-
   developer/table/3541286?hl=en&rd=1
   ```

2. Go to the Developer Play Console:

   ```
   play.google.com/apps/publish/
   ```

3. Sign in with your Google account and enter the following information:

Developer Name	_____
	Will appear to users under the name of your application
Email Address	_____
Website URL	_____
Phone Number	_____
	Include plus sign, country code and area code. For example, +1-650-253-0000.
Email Updates	☐ Contact me occasionally about development and Google Play opportunities.

4. Read and accept the **Google Play Developer Distribution Agreement**.

5. Pay the USD 25 with Google Checkout, creating an account if necessary, and that's it, you are now a registered Google Developer.

If you intend to make your apps available worldwide, then it is always worth checking the Supported Locations page, as it changes regularly. The only thing left to do is upload our app, which we will do next.

Publishing an app on the Google Play store

Uploading and publishing our apps to the Play store is done through the **Developer Console**. As you will see, there is a lot of information and promotional material that we could provide about our app during this process. Providing you have followed the previous steps in this chapter and have a release-ready signed `.apk` file, then complete the following instructions to publish it. Alternatively, you may just want to have a look at what is involved at this point and what form the promotional material will take. In this case, ensure you have the following four images and a signed APK and select **Save Draft** at the end rather than **Publish app**:

1. At least two screenshots of your app. These must not have any side that is shorter than 320 px or longer than 3840 px.

2. If you want your app to be visible on the Play store to users searching for apps designed for tablets, then you should prepare at least one 7 " and one 10 " screenshot.

3. A Hi-res icon image of 512 x 512 px.

4. A Feature Graphic of 1024 x 500 px.

With these images prepared, and a signed .apk, we have all we need to start. Decide how much, if anything, you wish to charge for the app and then follow these instructions:

1. Open your Developer Console.

2. Supply a **Title** and click on the **Upload APK** button.

3. Click on the **Upload your first APK to Production**.

4. Locate your signed `app-release.apk` file. It will be in `AndroidStudioProjects\YourApp\app`.

5. Drag and drop this into the space suggested.

6. When this is completed, you will be taken to the application page.

7. Work your way through the top four sections:

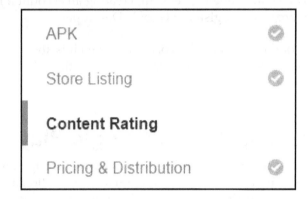

8. Complete all required fields until the Publish app button becomes clickable.
9. If you need help, the **Why can't I publish?** link above the button will list uncompleted compulsory fields.
10. When all the required fields are completed, click on the **Publish app** (or **Save draft**) button at the top of the page.
11. Congratulations! You are now a published Android developer.

We now know how to publish our apps on the Play store. There are, of course, many other app markets, and they all have their own uploading procedures. Google Play, however, provides the widest possible audience and is the obvious choice for publication.

Even though the Play store makes the ideal market place, it is still well worth looking at two alternative methods of distributing.

Distributing by e-mail and on websites

The first of these two methods is as simple as it sounds. If you attach the APK to an e-mail and it's opened on an Android device, the user will be invited to install the app when the attachment is opened. On more recent devices, they will be able to tap an install button directly in the e-mail.

> For both these methods, your users will have to allow the installation of unknown sources in the security settings of their devices.

Distributing your app from your website is almost as simple as e-mailing it. All you need to do is host the APK file on your site somewhere and provide a download link along the lines of the following:

```
<a href="download_button.jpg" download="your_apk">.
```

When browsing your site from and Android device, a tap on your link will install your app on their device.

> Distribution by e-mail provides no protection against piracy and should only be used with this in mind. The other methods are as secure as we could hope, but if you would like to take extra measures, then Google offer a **Licensing Service** which can be found at developer.android.com/google/play/licensing.

Whether we have released a paid app or a free one, we want to be able to reach as many users as possible. Google provide several tools to help us with this, as well as ways to monetize our apps, as we shall see next.

Promoting and monetizing apps

Very few apps become successful without first being well promoted. There are countless ways to do this and you will, no doubt, be well ahead of the curve on how to promote your products. To help you reach a wider audience, Google provides some handy tools to assist with promotion.

After looking at promotion tools, we will explore two ways to make money from our app: in-app payments and advertising.

Promoting an app

There are two very simple methods, provided by Google, to help steer people towards our products on the Play store: links from both websites and our apps, and the **Google Play Badge**, which provides official branding to our links.

We can add links to both individual apps and our publisher page, where all our apps can be browsed, and we can include these links in our apps as well as our websites:

- To include a link to a specific app's page in the Play store, use the full package name, as found in the Manifest, in the following format:

```
http://play.google.com/store/apps/details?id=com.full.package.name
```

- To Include this within an Android app, use this:

```
market://details?id= com.full.package.name
```

- If you want a link to your publisher page and a list of all your products, use this:

```
http://play.google.com/store/search?q=pub:my publisher name
```

- Make the same changes as before when linking from an app:

```
Market://search?q=pub:my publisher name
```

- To link to a specific search result, use this:

```
search?q=my search query&c=apps.
```

- To use an official Google Badge as your link, replace one of the preceding elements with the highlighted HTML here:

```
<a href="https://play.google.com/store/search?q=pub: publisher
name">
<img alt="Get it on Google Play"
src="https://developer.android.com/images/brand/en_generic_rgb_wo_60.png"
/>
</a>
```

The Badge comes in two sizes, `60.png` and `45.png`, and two styles, Android app on Google Play and Get it on Google Play. Simply change the relevant code to select the Badge that best suits your purpose:

With our app published and with well-placed links to our Play store page, it is now time to consider how we can profit from the inevitable downloads, and so we come to how to monetize and Android app.

Monetizing an app

There are many ways to make money from an app, but two of the most popular and effective are **in-app billing** and **advertising**. In-app billing can become quite involved and perhaps deserves an entire chapter to itself. Here, we will see how to build an effective template that you can use as a foundation for an in-app product you might develop. It will include all the libraries and packages needed, along with some very useful helper classes.

Including Google AdMob advertisements in our apps is, in contrast, a very familiar process to us by now. An ad is in effect just another View and can be identified and referenced just like any other Android widget. The final exercise of this chapter, and indeed the book, will be constructing a simple working AdMob demo. First, though, let's take a look at in-app billing.

In-app billing

There is a large number of products that users can purchase from within an app, from upgrades and unlockables to in-game objects and currencies, and it would certainly provide a payment option for the sandwich builder app we developed earlier in the book.

Whatever the user is buying, the Google checkout process ensures they will pay in the same way as they pay for other Play store products. From the developer's point of view, each purchase will boil down to responding to the click of a button. We will need to install the Google Play Billing Library, and add an AIDL file and some helper classes to our project. Here is how:

1. Start a new Android project or open one you want to add in-app billing to.
2. Open the SDK Manager.
3. Under Extras, make sure you have the Google Play Billing Library installed.
4. Open the manifest and apply this permission:

   ```
   <uses-permission
       android:name="com.android.vending.BILLING" />
   ```

5. In the Project pane, right-click on app and select **New | Folder | AIDL Folder**.
6. From this AIDL folder, create a **New | Package**, and call it com.android.vending.billing.
7. Locate and copy the `IinAppBillingService.aidl` file in the `sdk\extras\google\play_billing` directory.
8. Paste the file into the `com.android.vending.billing` package.
9. Create a **New | Package** in the Java folder called `com.`**your.package.name**`.util` and click on **Finish**.
10. From the `play_billing` directory, locate and open the `TrivialDrive\src\com\example\android\trivialdrivesample\util` folder.
11. Copy the nine Java files into the util package you just created.

You now have a working template for any app you wish to include in-app purchasing in. Alternatively, you can complete the preceding steps on a project where you have already developed your in-app products. Either way, you will no doubt be taking advantage of the `IabHelper class`, which vastly simplifies coding, providing listeners for every step of the purchasing process. Documentation can be found here:

developer.android.com/google/play/billing/billing_reference

> Before you can start to implement in-app purchases, you will need to secure a **License Key** for your app. This can be found in the app's details in your developer console.

Paid apps and in-app products are just two ways to make money from an app, and many people choose another, and often lucrative, route for monetizing their work through advertising. **Google AdMob** allows for a great deal of flexibility and a familiar programming interface, as we shall see next.

Including an advertisement

There are many ways that we can earn money from advertising, but AdMob provides one of the easiest. Not only does the service allow you to select what types of product you wish to advertise, but it also provides great analytical tools and seamless payment into your Checkout account.

On top of this, an **AdView** can be treated programmatically in a way that is almost identical to the methods we are used to and familiar with, as we shall see in this final exercise, where we will develop a simple app with a demo banner AdMob ad.

Before you start this exercise, you will need to have signed up for an AdMob account at google.com/admob:

1. Open a project you want to test ads on or start a new Android project.
2. Make sure you have the Google Repository installed with the SDK Manager.
3. In the `build.gradle` file, add this dependency:

```
compile 'com.google.android.gms:play-services:7.0.+'
```

4. Rebuild the project.
5. In the manifest, set these two permissions:

```
<uses-permission
    android:name="android.permission.INTERNET" />
<uses-permission android:name="android.permission.ACCESS_NETWORK_STATE"
/>
```

6. Within the application node, add this `meta-data` tag:

```
<meta-data
    android:name="com.google.android.gms.version"
    android:value="@integer/google_play_services_version" />
```

7. Include this second Activity in the manifest:

```
<activity
    android:name="com.google.android.gms.ads.AdActivity"
    android:configChanges=

"keyboard|keyboardHidden|orientation|screenLayout|uiMode|screenSize|smalles
tScreenSize"
    android:theme="@android:style/Theme.Translucent" />
```

8. Add the following string to the `res/values/strings.xml` file:

```
<string name="ad_id">ca-app-pub-3940256099942544/6300978111</string>
```

9. Open the `main_activity.xml` layout file.

10. Add this second namespace to the root layout:

```
xmlns:ads="http://schemas.android.com/apk/res-auto"
```

11. Add this `AdView` under the `TextView`:

```
<com.google.android.gms.ads.AdView
    android:id="@+id/ad_view"
    android:layout_width="match_parent"
    android:layout_height="wrap_content"
    android:layout_alignParentBottom="true"
    android:layout_centerHorizontal="true"
    ads:adSize="BANNER"
    ads:adUnitId="@string/ad_id"></com.google.android.gms.ads.AdView>
```

12. In the `onCreate()` method of `MainActivity,` insert these lines:

```
AdView adView = (AdView) findViewById(R.id.ad_view);
AdRequest adRequest = new AdRequest.Builder()
        .addTestDevice(AdRequest.DEVICE_ID_EMULATOR)
        .build();

adView.loadAd(adRequest);
```

13. Now test the app on a device.

More or less everything we did here resembles the way that we would program any other element, with one or two exceptions. The use of the ACCESS_NETWORK_STATE permission is not strictly necessary; it is used here to check for a connection prior to requesting an ad.

Any Activity that displays an ad will require a separate ID and be declared in the manifest. The ID supplied here is for testing purposes only as it is not allowed to use live IDs for testing purposes. There are only six classes in the `android.gms.ads` package, and documentation for all of them can be found at `developer.android.com/reference/com/google/android/gms/ads/package-summary`.

AdMob ads come in two flavors, the banner that we saw here and the interstitial, or full screen. We only dealt with banner ads here, but interstitial ads are handled in a very similar manner. With knowledge of how to implement paid apps, in-app billing, and AdMob, we are now armed to reap the rewards of our hard work and make the very most of our apps.

Summary

This chapter has outlined the final stages of app development, and although these stages only make up a small proportion of the workload, they are essentially important and can make all the difference when it comes to the success of an application.

Throughout the book, we have relied heavily on support libraries to increase the number of devices our apps can run on, but here we saw how we can extend that range even further by dynamically determining the platform and running appropriate code accordingly. This process provided a lovely example of how design patterns can pervade all aspects of programming.

Once we have used these tools to extend our reach, we can further enhance our app's chances of success with prudent promotion and hopefully make our work pay, either directly by charging our users for the app or its features, or indirectly through hosting advertisements.

Throughout the book, we have looked at how design patterns can assist us in many aspects of development, but it is a way of thinking that makes patterns so useful rather than any individual pattern itself. Design patterns provide an approach to problem solving and a clear path to solutions. It is an approach designed to guide us to new creative solutions, and design patterns should not be seen as written in stone but more as a guide, and any pattern can be modified and altered to better suit its purpose.

The pattern and samples in this book are not designed to be cut and pasted into other project, but rather as a methodology to help us find the most elegant solutions to our own original situations. If this book has done its job, then the patterns you go on to design will not be the ones outlined here but entirely new and original creations of your own.

Index